This work represents an impressive, well-thought-out, concise and simplified approach to structural design. The break-out of simplified load diagrams, appropriate formulae, their application, and material studies, allow for ease of understanding and retention. This book should be owned by every architectural student, intern, and practicing architect.

**—Wendell C. Edwards, RA, PhD, NCARB, Associate Professor,
New York City College of Technology /
The City University of New York**

Using a clear, well thought-out format, this book provides a practical guide for architects to understand the many aspects of structural design. It incorporates theory, analysis of structural forces, a guide to the properties of wood, steel and concrete, and a review of design considerations to be addressed. Applying their extensive practical experience designing buildings and structures, Professors Bedi and Dabby use multiple case studies to illustrate the process, carefully explaining each step of the structural design process. The multiple clear diagrams and illustrative sketches, carefully selected photographs, and very useful reference charts contribute to making this an extremely useful and much needed book.

**—Professor Tim Maldonado FARA / Architect,
Former Dean of the School of Technology & Design,
New York City College of Technology / The City University of New York**

Structure for Architects offers a comprehensive overview on the principles of design for three major structural materials: steel, wood and reinforced concrete. The content is presented in a clear, intuitive and visual manner, and includes valuable case studies that help illustrate and enhance the concepts. An excellent reference and teaching tool for students of architecture.

**—Ivan Markov, PhD, Structural Engineer,
CCNY Spitzer School of Architecture /
The City University of New York**

The authors have succeeded in presenting an excellent resource for understanding the fundamental principles of structural engineering that serves to benefit both practicing architects and architectural students as well as providing a refresher for more experienced engineers. As technological advancements have moved our profession to an increased reliance on computerized solutions to complex structural problems, it is important to maintain touch with basic engineering concepts as applied to the design of steel, wood, and reinforced concrete framing. This reference book provides a firm platform for achievement of that result.

**—Thomas J. Michon, PE, Senior Structural Engineer,
Dormitory Authority of the State of New York (Retired)**

Structure for Architects: A Case Study in Steel, Wood, and Reinforced Concrete Design is a descriptive and well-organized text that outlines the basics of structural design. I was pleasantly surprised by the impact of the illustrations and their ability to reinforce the content of each chapter. The case studies show the application of theory and serve as a guide for general design principles. It is a fantastic tool for introducing architects to the world of structural design.

**—Paul Senica, Structural Engineer, Murray Engineering PC,
Engineering Adjunct Faculty, New York Institute of Technology**

This volume builds on the authors' previous *Primer* to provide a valuable and engaging set of texts for teaching structures for architecture students. The endeavor has been informed by years of classroom experience, a passion for the subject, and an intuitive understanding of the power of visual representations to convey structural principles. The variety and clarity of the many illustrations—photographs, diagrams, rendered axonometrics, and especially the beautiful and tactile hand sketching—is a great strength of the book. The authors know how to exploit the curiosity of the visual thinker and invite the type of exploration that leads to a genuine understanding.

—Shelley E. Smith, PhD, RA, Professor and Former Chair,
Department of Architectural Technology,
New York City College of Technology /
The City University of New York

The case studies illustrate a workflow and parametric thinking that integrates design, structure, and materials. It is a valuable tool for architects to analyze and solve problems methodically.

—**Sanjive Vaidya, RA, Chair of the Department of Architectural Technology,**
New York City College of Technology /
The City University of New York

Structure for Architects

Structure for Architects: A Case Study in Steel, Wood, and Reinforced Concrete Design is a sequel to the authors' first text, *Structure for Architects: A Primer*, emphasizing the conceptual understanding of structural design in simple language and terms. This book focuses on structural principles applied to the design of typical structural members—a beam, a girder, and a column—in a diagrammatic frame building. Through the application of a single Case Study across three key materials, the book illustrates the theory, principles, and process of structural design. The Case Study progresses step-by-step for each material, from determining tributary areas and loads through a member's selection and design.

The book addresses the frequent disparity between the way architects and engineers perceive and process information, with engineers focusing on technical aspects and architects focusing on visual concepts. *Structure for Architects: A Case Study in Steel, Wood, and Reinforced Concrete Design* presents readers with an understanding of fundamental engineering principles through a uniquely thematic Case Study. Focusing on the conceptual understanding of structural design, this book will be of interest to architecture students and professionals looking to understand the application of structural principles in relation to steel, wood, and concrete design.

Ashwani Bedi is a Professional Engineer and Deputy Director of Engineering with the NYC Department of Design and Construction. He has over 30 years of experience in the public and private sectors, and is currently Adjunct Associate Professor at the New York City College of Technology, where he teaches Structures courses. He is also the co-author of *Structure for Architects: A Primer*.

Ramsey Dabby is a Registered Architect having served as Chief Architect for the NYC Fire Department, and previously as Director of Engineering and Construction for the Hertz Corporation. Now retired with over 40 years' experience in the public and private sectors, he currently teaches as Adjunct Faculty in the Architectural Departments of the New York City College of Technology, the City College of New York, and the New York Institute of Technology. He is also the co-author of *Structure for Architects: A Primer*.

Structure for Architects

A CASE STUDY IN STEEL, WOOD, AND REINFORCED CONCRETE DESIGN

Ashwani Bedi *and*
Ramsey Dabby

Routledge
Taylor & Francis Group
NEW YORK AND LONDON

First published 2020
by Routledge
52 Vanderbilt Avenue, New York, NY 10017

and by Routledge
2 Park Square, Milton Park, Abingdon, Oxon, OX14 4RN

Routledge is an imprint of the Taylor & Francis Group, an informa business

Library of Congress Cataloging-in-Publication Data
Names: Bedi, Ashwani, author. | Dabby, Ramsey, author.
Title: Structure for architects : a case study in steel, wood, and reinforced concrete design / Ashwani Bedi and Ramsey Dabby.
Description: New York : Routledge, 2019. | Includes bibliographical references and index.
Identifiers: LCCN 2018057508 | ISBN 9781138554375 (hardback) | ISBN 9781138554382 (pbk.) | ISBN 9781315122014 (e-book)
Subjects: LCSH: Structural engineering. | Girders. | Columns.
Classification: LCC TH845 .B43 2019 | DDC 324.102/472—dc23
LC record available at https://lccn.loc.gov/2018057508

ISBN: 978-1-138-55437-5 (hbk)
ISBN: 978-1-138-55438-2 (pbk)
ISBN: 978-1-315-12201-4 (ebk)

Typeset in News Gothic MT
by Apex CoVantage, LLC

Contents

Preface xi
Image Credits xiii

1 Introduction to Structural Design 1

1.1 Removing a Bearing Wall 2

2 Structural Design Methodologies 4

2.1 Building Codes and Industry Organizations 4
2.2 Loads 6
2.3 Safety Factors 7
2.4 Load Combinations 7
2.5 Design Methodologies 8

3 Stress, Strain, and Material Behavior 14

3.1 Stress, Strain, and Material Properties 14
3.2 Stress-Strain Curves for Steel, Wood, and Concrete 16
3.3 Elastic/Plastic Material Behavior—An Analogy 22
3.4 Elastic/Plastic Analogy Applied to a Beam 25

3.5 Section Modulus 27
3.6 Getting Started—Case Study 29

4 Case Study Introduction 31
4.1 Load Combinations 32
4.2 Designing for Beams and Girders 35
4.3 Designing for Columns 41

5 Understanding Steel 44
5.1 Manufacture and Materials 44
5.2 General Design Considerations 48
5.3 Design Considerations for Beams 49
5.4 Design Considerations for Columns 51

6 Design in Steel—Case Study 54
6.1 Assumptions 55
Case Study—Design in Steel (ASD) 57
6.2.ASD Beam 3 57
6.3.ASD Girder B 63
6.4.ASD Column B2 67
Case Study—Design in Steel (LRFD) 71
6.2.LRFD Beam 3 71
6.3.LRFD Girder B 76
6.4.LRFD Column B2 79
ASD/LRFD Discussion 82

7 Understanding Wood 83
7.1 Sawn Lumber—Manufacture and Properties 83
7.2 Sawn Lumber—Design Considerations for Beams 89

7.3	Sawn Lumber—Design Considerations for Columns	91
7.4	Engineered Lumber—Manufacture and Products	92
7.5	Engineered Lumber—Design Considerations	95
8	**Design in Sawn Wood—Case Study**	**96**
8.1	Assumptions	97
	Case Study—Design in Sawn Wood	100
8.2	Beam 3	100
8.3	Girder B	105
8.4	Column B2	109
9	**Design in Engineered Wood—Case Study**	**113**
9.1	Assumptions	114
	Case Study—Design in Engineered Wood	116
9.2	A Typical Joist	116
9.3	Girder B	119
9.4	Column B2	121
10	**Understanding Reinforced Concrete**	**128**
10.1	Materials and Manufacture	128
10.2	Structural Considerations	132
10.3	Design Considerations for Beams	133
10.4	Design Considerations for Columns	148
11	**Design in Reinforced Concrete—Case Study**	**152**
11.1	Assumptions	153
	Case Study—Design in Reinforced Concrete	157
11.2	Beam 3	157

11.3 Girder B 164
11.4 Column B2 171

12 In Closing 175

Glossary 177
Appendix 1: The AISC Steel Construction Manual 189
Appendix 2: The National Design Specification (NDS) Package for Wood Construction 205
Appendix 3: ACI 318—Building Code Requirements for Structural Concrete 213
Appendix 4: Beam Diagrams and Formulae 217
Index 223

Cover Illustration Courtesy of:
Artur Nesterenko Alexandrovich
Archillusion Design and Development
www.archillusion.com

Preface

The underlying idea of *Structure for Architects* arose from our classroom experiences in the Architectural Technology Department of the New York City College of Technology. We perceived the need for texts that presented the topic of structures to architectural students, in a more intuitive and visual manner—and so the concept was born.

Our first text, *Structure for Architects: A Primer*, presented basic structural principles laying the groundwork for more advanced studies. In this text, *Structure for Architects: A Case Study in Steel, Wood, and Reinforced Concrete Design*, those principles are applied to the structural design of typical members—a beam, a girder, and a column—in a diagrammatic frame building. For each material, the Case Study progresses in a consistent step-by-step manner—from determining tributary areas and loads, free-body, shear and moment diagrams, through the selection and design of members. We've assumed the Reader is familiar with basic structural concepts, and provide only brief recaps where appropriate.

While the information in this text is generally available from a variety of sources, our goal was to present material in a single source explaining theory and principles, hopefully providing the Reader with an intuitive understanding of the processes and formulae. Of note are the chapters titled "Understanding" that precede the Case Studies, in which pertinent properties and design information for each material are presented.

In retrospect we once again realize the very considerable amount of time, patience, and dedication needed in an effort such as this. We hope to have done it justice. We wish to acknowledge several people for their contributions to us. Any shortcomings, oversights, or errors that may have occurred are the sole responsibility of the authors. In no particular order, we sincerely thank:

- Professor Robert Bunnell, PE, adjunct faculty at Kent State University, College of Architecture and Environmental Design, and the University of Akron,

Department of Civil Engineering. Professor Bunnell's exhaustive and thoroughly professional review of the manuscript was invaluable in shedding light on numerous technical and educational concepts.

- Professor Sanjive Vaidya, Chair of the Architectural Technology Department at the NYC College of Technology, for his time and thoughts on the book's conceptual framework.
- Professor Tim Maldonado, former Dean of the School of Architecture and Engineering Technology at the NYC College of Technology, for his perspective, encouragement, and guidance—and for his vision to pair an engineer and architect in the teaching of Structures.
- Professor Shelley Smith, former Chair of the Architectural Technology Department at the NYC College of Technology, for her leadership and dedication to excellence, independent thought, and creativity.
- Tom Michon, our dear friend and accomplished structural engineer, for his thorough technical reviews and support.
- Richard Gilbert, Principal of BlueShore Engineering, for his practical yet scholarly opinion on a number of topics.
- Christian Martos, PE, Principal of Martos Engineering and adjunct faculty at CCNY, for his easy-to-understand descriptions of NDS Adjustment Factors.
- Art Lewis, of Weyerhaeuser Corporation, for his support on the engineered wood sections.
- Zoe van der Meulen and Liz Connor, of Unalam, manufacturers of quality glulam products.
- Havard Vasshaug and Bad Monkeys for their creative and technical genius, and their approach to the dissemination and sharing of knowledge.
- Former student Moe Goldstein for his review from a student's perspective.
- Artur Nesterenko Alexandrovich for his cover illustration.

And finally our families who endured our consumed weekends, holidays, and spare time at their expense. Here's to them, especially Kiran and Louise Harris, who patiently encouraged the completion of this work—and to Barbara who is still missed every day.

We welcome and appreciate any comments from the Reader.

Ashwani Bedi, PE (akbedi@gmail.com)

Ramsey Dabby, RA (rdabby@optonline.net)

This text cannot, and makes no attempt to, cover the many complexities that structural engineers routinely encounter. Professional engineering expertise should of course always be sought as needed.

IMAGE CREDITS

The authors gratefully acknowledge the following for permissions to use their images (Figure numbers):

American Concrete Institute (Appendix 3 Tables)
American Institute of Steel Construction (Appendix 1 Tables)
American Wood Council (Appendix 2 Tables)
Andrew Haase (4.5)
Buckland Timber (7.14)
Capital Steel & Wire (5.7)
Chicago Architecture Center, George N. Bernard (2.2)
Chirag N. Patel, PhD (10.27)
Clark Hyland, PhD, Hyland Consulting Engineering (10.36)
David Harrington (11.2)
DoITPoMS (www.doitpoms.ac.uk/tlplib/wood/printall.php) (7.6)
Havard Vasshaug and Bad Monkeys (10.15)
InterNACHI (10.8)
Nina Edwards Anker, NEA Studio; and Unalam (7.11)
Owens Supply (10.11)
Paolo Rosselli, Studio di Fotografia (1.2)
Portland Cement Association (10.6)
Public Domain (2.1, 2.5, 5.2, 5.3a–b, 5.4–5.6, 7.1)
Robor Steel (5.10)
StructureCraft Builders (7.17)
Swinerton Builders (7.15)
The Liro Group (10.38)
Thyssenkrupp Elevators (2.4)
US Geological Survey (2.6)
WCLIB (7.5)
Weyerhaeuser (7.12, 7.13, Tables 9.1–9.5)

1 Introduction to Structural Design

One of the earliest and most important engineering decisions on a project is to select a structural system that is compatible and consistent with the architectural intent. On some projects, the structural system may conform to the architectural expression and simply be a concealed means of support (Figure 1.1). On other projects, the structural system may be the actual architectural expression and be celebrated for itself (Figure 1.2). In any case, the selection of an appropriate structural system is a matter of discussion and close coordination between architect and structural engineer, keeping in mind the best interests of the project.

Whatever the structural system and however it may be expressed, the structure must be designed to satisfy the conditions of stability, strength, serviceability, economy, and sustainability—not only as a whole, but also for its individual components.

Figure 1.1 Apartment Building, Fort Lee - Structure as Concealed Support

Figure 1.2 Bach de Roda Bridge, Barcelona - Structure as Architectural Form

With a good general understanding of how structural members behave under load, the Reader is ready to undertake their design.

> *The structural design of a member simply means the selection of an appropriate material and cross-sectional shape to safely and economically resist the load demands, to which the member will be subjected.*

1.1 REMOVING A BEARING WALL

Let's use a simple example to get a sense of this statement.

Say you're renovating the basement of a house and want to open up the room, column-free, by removing a bearing wall that runs down the middle of the space (Figure 1.3). Since the bearing wall is supporting the first floor joists above, you know that with the bearing wall removed, you'll have to support the loads from the joists with a new girder. Since the floor-to-floor height is limited, headroom clearance is a concern.

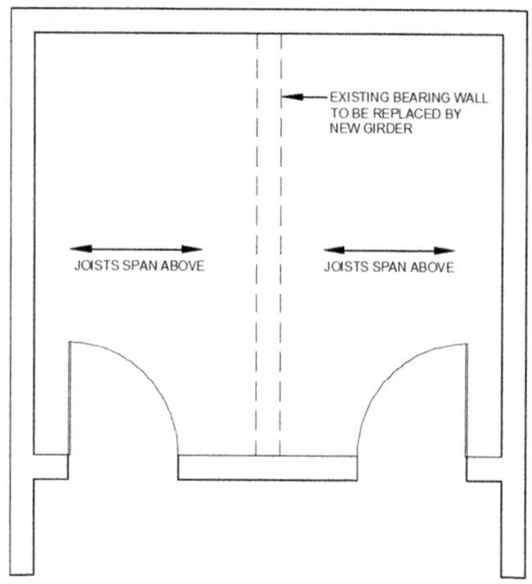

Figure 1.3 Basement Plan

Option 1

Your first instinct, for reasons of structural efficiency, practicality, and cost may be to design a deep and narrow girder in wood—but you're concerned that the girder may be too deep for sufficient headroom.

Option 2

You realize however, that you can increase headroom by compromising structural efficiency and making the wood girder shallower and wider.

Option 3

It now occurs to you to design the girder in steel, knowing that steel has significantly greater strength than wood, thereby making it feasible to design a shallow steel girder providing the needed headroom—but is likely that this will be a more expensive solution (Figure 1.4).

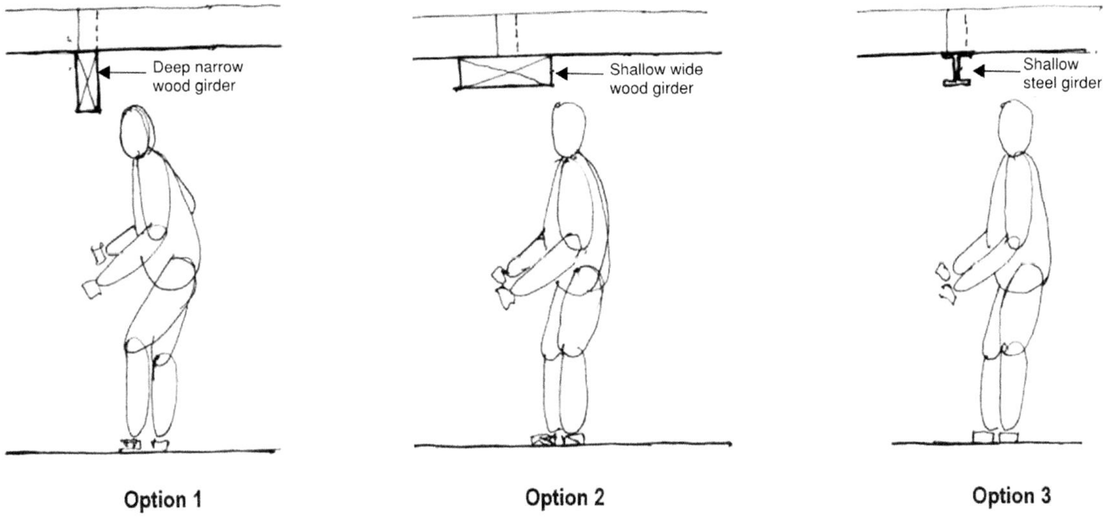

Figure 1.4 Girder Options

Once you calculate the loads on the girder, and determine the moment and shear that the girder must resist, you'll have the basic information needed to design the beam—but which option is the right choice?

This example illustrates the fundamental challenge of design—to use judgment and experience to reach a decision when there are multiple options that can satisfy the design intent. The decision of whether to design the girder in wood or steel is a matter of preference of the designer based on the availability and economy of the materials, and ease of construction—all of which may vary for any particular situation. In the end, you as the designer must evaluate the key properties of a member that will provide the strength to safely and economically carry the anticipated loads. Specifically, you must decide upon a member's material and cross-sectional shape.

2 Structural Design Methodologies

2.1 BUILDING CODES AND INDUSTRY ORGANIZATIONS

Building codes are a set of rules that provide the public with a minimum acceptable level of performance for buildings and other structures. Their main purpose is to protect public health, safety, and welfare as they relate to a structure's construction and occupancy.

The advent of building codes is generally traced to Hammurabi's Code during the Babylonian Empire around 1700 BC (Figure 2.1). The earliest known building codes in the United States were established in the 1600s, but it wasn't until the 1800s that larger US cities began adopting building codes, spurred largely by catastrophes like the Great Chicago Fire of 1871 (Figure 2.2). These early versions have developed into the building codes in use today.

Until recently, the United States followed various local and regional codes such as the BOCA National Building Code, the Uniform Building Code, and the Standard Building Code. Variations within these codes, along with the desire to have a unified

Figure 2.1 Hammurabi's Code

Figure 2.2 1871 Chicago Fire

building code, gave birth to the International Code Council (ICC) in 1994 that was comprised of officials representing all three codes. The ICC published the first edition of the International Building Code (IBC) in 2000, with updated revisions in subsequent years. Based on the IBC, cities and states develop and adopt their own 'governing' codes that reflect their own specific conditions and requirements.

Along with the development of building codes, advances in the science and behavior of various construction materials (such as steel, wood, and concrete) led to the formation of 'Societies', 'Institutes', and 'Councils' comprised of industry manufacturers, academic and practicing professionals. These organizations evolved into technical authorities in their fields, performing research, providing guidelines, and becoming references for the various building codes. The American Institute of Steel Construction (AISC), the American Wood Council (AWC), and the American Concrete Institute (ACI) are the recognized respective authorities governing the design, construction, and use of steel, wood, and concrete. These organizations produce various technical publications that are adopted by reference in various codes, and accepted as industry standards for the proper and safe use of these materials. These publications include:

- the AISC Steel Construction Manual (see Appendix 1)
- the National Design Specification (NDS) Package for Wood Construction (see Appendix 2)
- Building Code Requirements for Structural Concrete—ACI 318 (see Appendix 3)

Other important organizations such as the American Society of Civil Engineers (ASCE), the American Society for Testing and Materials (ASTM), and the American National Standards Institute (ANSI) are also recognized authorities that support the civil engineering profession by providing technical publications and reports on technical matters.

Figure 2.3 gives a general sense of the interaction of codes and various technical organizations.

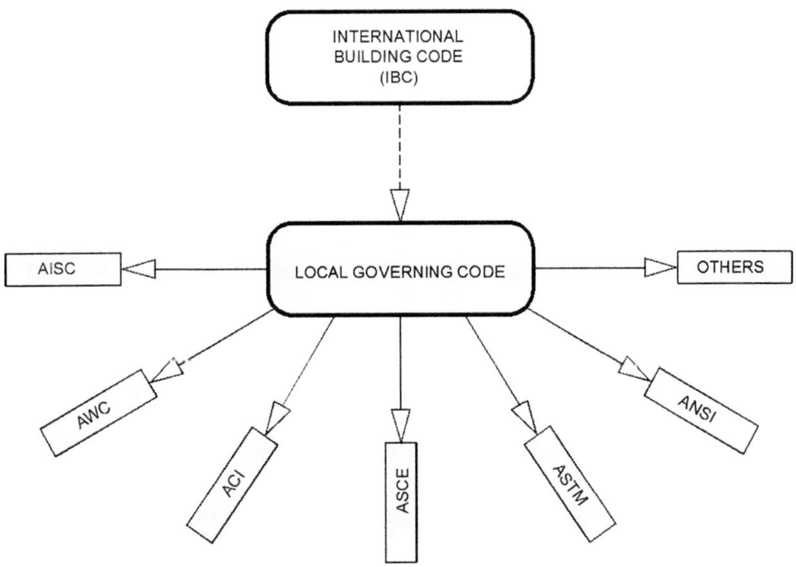

Figure 2.3 Codes and Technical Organizations

2.2 LOADS

There are many types of loads on a structure, which can be broadly classified as dead, live, and environmental.

Dead Loads

Dead loads are fixed and of constant magnitude. They include predictable factors such as floor and wall construction, partitions, ceiling and floor finishes, mechanical/plumbing/electrical, and other fixed equipment (Figure 2.4). The weights of construction materials are available from various sources, including the AISC Manual.

Figure 2.4 Dead Load—elevator machinery

Live Loads

Live loads are temporary, transient, and variable. They include things such as people, furniture and equipment, vehicles, and stored goods. Live loads are typically measured in pounds per square foot (psf) and are given in local building codes according to the specific use and occupancy of a building (Figure 2.5).

Figure 2.5 Live Load—runners on a bridge

Environmental Loads

Environmental loads are really live loads, but those resulting from the effects of natural phenomena. They vary based upon the geographic location of the structure, and include things such as rain, ice, snow, wind, and seismic activity (earthquakes) (Figure 2.6).

Figure 2.6 Earthquake Damage

In the analysis of forces on a structure, loads are often also categorized as static and dynamic.

Static refers to gravity-type forces such as the dead loads previously mentioned and some of the live and environmental loads such as human occupancy, moveable furniture and equipment, stored goods, and snow.

Dynamic refers to forces that are sudden over limited periods of time such as wind, seismic activity, impact, machinery vibrations, moving vehicles, and elevators.

Ultimately all loads are treated as forces and/or moments that act vertically or horizontally (Figure 2.7).

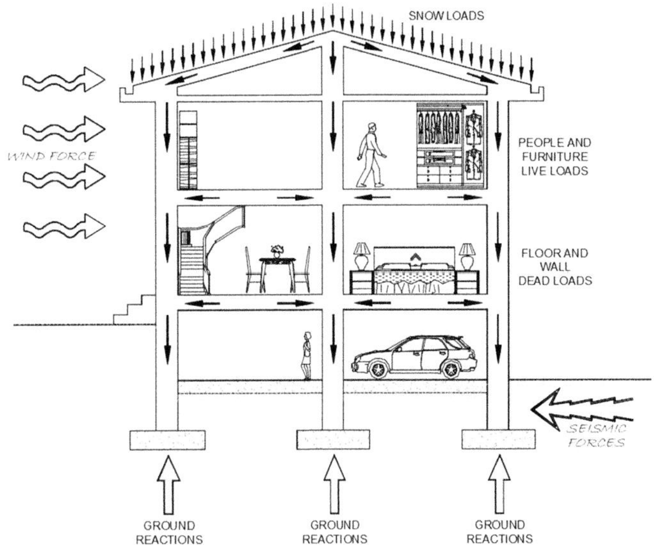

Figure 2.7 Loads on a Structure

2.3 SAFETY FACTORS

Uncertainties are inherent in the design and construction process since it is not possible to calculate with absolute accuracy the strength of materials, the loads, and the varying quality of construction. Load calculations performed by designers can only be an estimate of the loads to which a structure will be subjected. Codes allow for uncertainties by providing for margins of safety in the form of *safety factors*. Safety factors provide a degree of reliability against failure by allowing for unavoidable variations between estimated and actual conditions. Safety factors are applied to a material's strength (reducing it) and/or to the loads (generally increasing them).

The actual loads on a structure, without the application of safety factors, are referred to as *service loads*.

2.4 LOAD COMBINATIONS

While several types of loads (dead, live, snow, rain, ice, wind, seismic activity, etc.) can act on a structure, it's highly unlikely they would all act simultaneously and/or at full magnitude. For example, it is extremely improbable that live loads would be at their maximum with a major snowstorm, hurricane, and earthquake all occurring at the same time. On the other hand, the design must account for the load combination that has the most critical effect. So the question becomes how to combine loads reasonably.

The IBC, ASCE 7 (Minimum Design Loads for Buildings and Other Structures), and various other applicable building codes, prescribe *load combinations* to determine the appropriate loading to use. These load combinations vary based upon the design methodology being used. Structures must be designed to resist the most critical effects from the load combinations.

In the load combinations, each service load is multiplied by a *load factor* reflecting the uncertainty of that particular load. With a load factor applied, the resulting load is called a *factored load*. The load combinations resulting in the most critical effect are then used to compute the moments, shears, and other forces in the structure.

In general, the load combination with highest value will govern. However some load combinations account for the possibility of overturning and uplift, and will govern in those situations. This is particularly applicable to tall structures where high lateral loads are present. The Reader is referred to more advanced texts for such situations. We'll see the application of load combinations in subsequent chapters.

2.5 DESIGN METHODOLOGIES

Design methodologies are recognized approaches to the design of structural members. Various design methodologies have evolved over time, and continue to do so as more research is done and knowledge gained. These methodologies and their terminologies may appear to overlap somewhat in meaning and approach, sometimes tending to cause a degree of confusion. In this chapter we'll examine the various methodologies to help bring them into focus, but for now simply note that they are separate and distinct approaches that must be consistently followed without mixing or matching parts and formulae.

Limit-State Design Principles—Strength and Serviceability

Design methodologies are based on *limit-state* design principles, which define the boundaries of structural usefulness in terms of *strength* and *serviceability*. Strength limit-states relate to a structure's load-carrying capability and safety. Serviceability limit-states relate to a structure's performance under normal service conditions—i.e., the conditions for which it has been designed.

Serviceability can be considered a measure of the 'comfort level' of the building inhabitants and, among other things, involves checking that beam deflections and vibrations are within allowable limits. Structures must be designed so that no applicable strength or serviceability limit-states are exceeded.

Terminology

The following terminology will be frequently used throughout the text:

Service loads	The loads on a structure, without the application of load factors.
Factored loads	The loads on a structure, after the application of load factors.
Design loads	The loads obtained from the governing load combinations.
Yield stress	The stress at which a material begins to exhibit excessive and permanent deformation, and can no longer perform as intended.

Actual stress	The stress in a structural member created by the design loads.
Allowable stress	The stress permitted in a structural member, after the application of safety factors.
Nominal strength	The strength of a structural member, before the application of safety factors.
Available strength	The strength of a structural member, after the application of safety factors.
Design strength	The Load and resistance Factor Design (LRFD) and Strength Design term for "available strength".
Allowable strength	The Allowable Strength Design (ASD) term for "available strength".
Required strength	The strength needed by a structural member to support the design loads.
Capacity	A term synonymous with "available strength".

Stress and Strength-Based Methodologies

Design methodologies are fundamentally either stress-based or strength-based.

A stress-based methodology involves assuring that a member's allowable stress is equal to or greater than the actual stress (Figure 2.8a).

A strength-based methodology involves assuring that a member's available strength is equal to or greater than the required strength (Figure 2.8b).

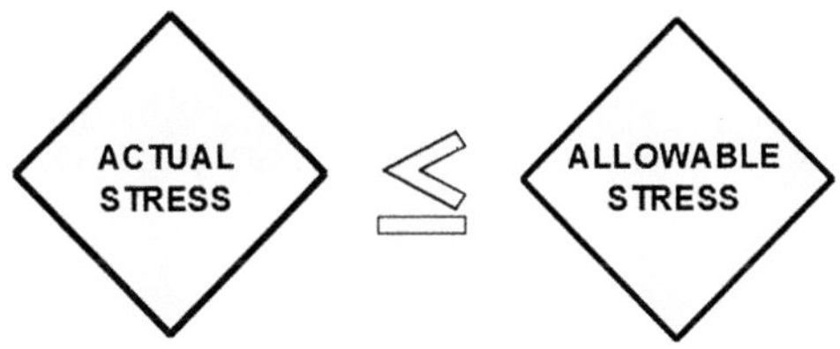

Figure 2.8a Stress-based Methodology

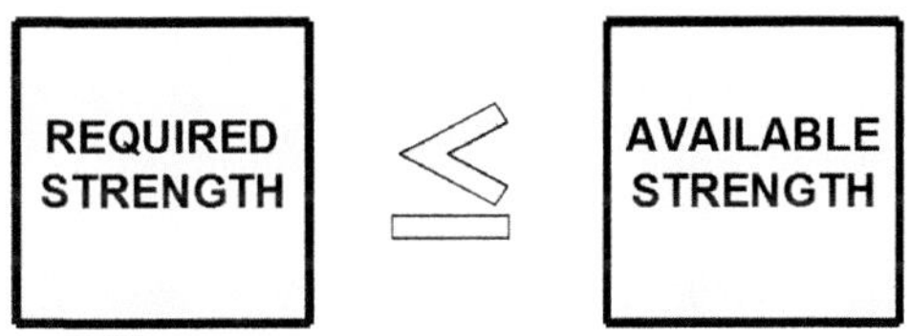

Figure 2.8b Strength-based Methodology

Allowable Stress Design

Background

Allowable Stress Design is the traditional approach dating from the early 1800s, in which safety factors are empirically derived. It is still in use today due to its conceptual simplicity and familiarity.

Concept

As a stress-based methodology, Allowable Stress Design involves assuring that a member's *actual stress is less than or equal to the allowable stress*. The allowable stress is determined by multiplying the material yield stress by a safety factor (less than 1), thereby reducing it. The actual stress is determined from the governing load combinations (Figure 2.9).

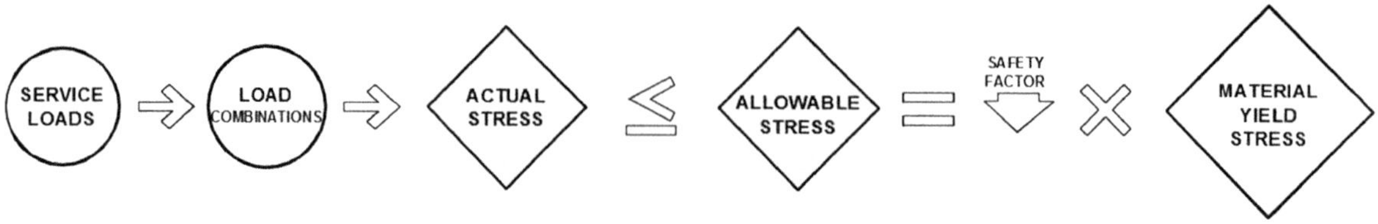

Figure 2.9 Allowable Stress Design

The relationship between allowable stress and actual stress can be expressed by:

$$f \leq F'$$

where:
f = actual stress
F′ = allowable stress

Use

- Allowable Stress Design is a recognized methodology for wood design.
- Allowable Stress Design is often referred to as Working Stress Design when used for certain types of concrete design.

Load and Resistance Factor Design (LRFD)

Background

Load and Resistance Factor Design was introduced in the 1980s by the AISC as an alternative, more rational, and exact method for steel design than Allowable Stress Design. This methodology resulted in potential economies in the design of members. With the development and adoption of LRFD, both LRFD and Allowable Stress Design became recognized methodologies for steel design until Allowable Strength Design superseded Allowable Stress Design.

Concept

As a strength-based methodology, LRFD involves assuring that a member's *available strength is equal to or greater than the required strength*. The available strength (termed *design strength* in LRFD) is determined by multiplying the material nominal strength by a resistance factor (less than 1), thereby reducing it. The required strength is determined from the governing load combinations (Figure 2.10).

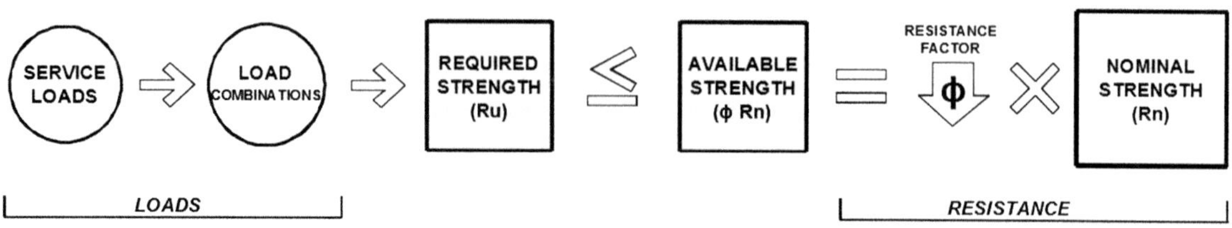

Figure 2.10 Load and Resistance Factor Design

The relationship between available strength and required strength is expressed by the LRFD general strength equation:

$\mathbf{R_u \leq \Phi R_n}$ where:

R_u = required strength
ΦR_n = available strength (design strength)
R_n = nominal strength
Φ = resistance factor

Use

- Load and Resistance Factor Design is a recognized methodology for steel design and wood design.

Allowable Strength Design (ASD)

Background

Allowable Strength Design was adopted in the 13th Edition of AISC Manual and superseded Allowable Stress Design as a recognized methodology for steel design. This change brought both steel design methodologies (ASD and LRFD) to a consistent strength-based platform.

Concept

As a strength-based methodology, ASD involves assuring that a member's *available strength is equal to or greater than the required strength*. The available strength (termed *allowable strength* in ASD) is determined by dividing the material nominal strength by a safety factor (greater than 1), thereby reducing it. The required strength is determined from the governing load combinations (Figure 2.11).

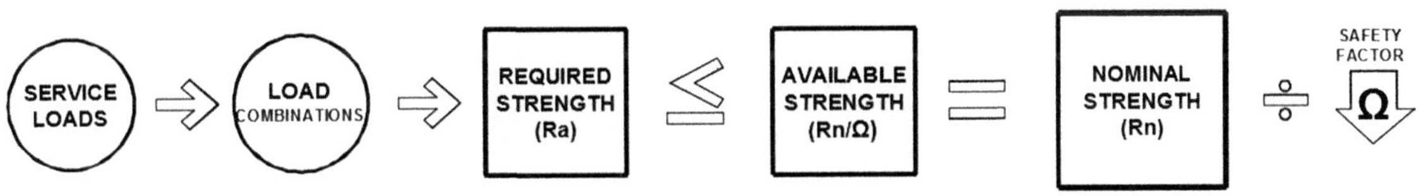

Figure 2.11 Allowable Strength Design (ASD)

The relationship between available strength and required strength is expressed by the ASD general strength equation:

$\mathbf{R_a \leq (R_n / \Omega)}$ where:

R_a – required strength
R_n / Ω = available strength (allowable strength)
R_n = nominal strength
Ω = safety factor

Use

- Allowable Strength Design, like Load and Resistance Factor Design, is also a recognized methodology for steel design.

Strength Design

Background

Prior to the 1950s, Working Stress Design was the predominant methodology for concrete. Ultimate Strength Design was then introduced to replace it. Since the 1971 ACI code, this has come to be known as *Strength Design*.

Concept

Strength Design is conceptually similar in approach to Load and Resistance Factor Design. As a strength-based methodology, Strength Design involves assuring that a member's *available strength is equal to or greater than the required strength*. The available strength (termed *design strength* in Strength Design) is determined by multiplying the material nominal strength by a strength reduction factor (less than 1), thereby reducing it. The required strength is determined from the governing load combinations (Figure 2.12).

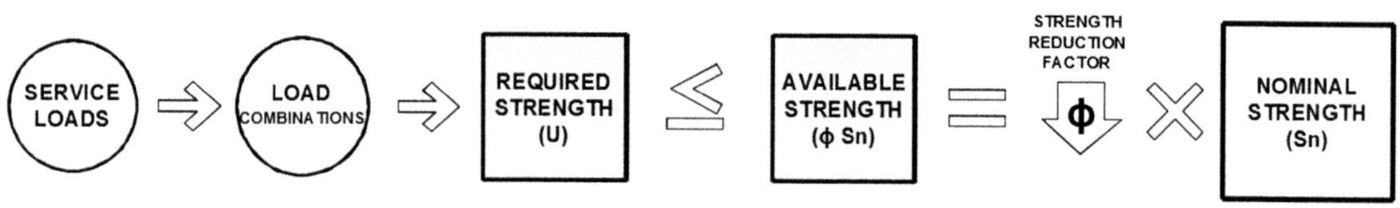

Figure 2.12 Strength Design

For the relationship between available strength and required strength, the ACI states "design strength ≥ required strength". This will be expressed by the following general strength equation:

$\Phi S_n \geq U$ where:

ΦS_n = available strength (design strength)
U = required strength
S_n = nominal strength
Φ = strength reduction factor

Use

- Strength Design is the predominant recognized methodology for reinforced concrete design. Strength Design is somewhat more involved than the other methodologies since it deals with two dissimilar materials (concrete and steel) working together.

Design Methodologies Applied

Figure 2.13 summarizes the recognized design methodologies for steel, wood, and reinforced concrete.

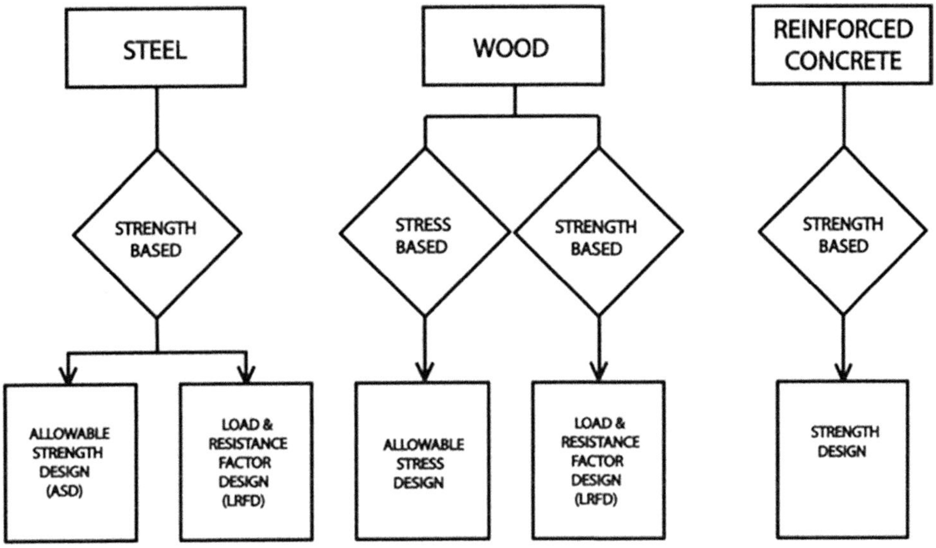

Figure 2.13 Materials and Design Methodologies

The unique behavior of a particular material under stress underlies the application of the various design methodologies. We'll examine the material behavior of steel, wood, and concrete in Chapter 3.

3

Stress, Strain, and Material Behavior

3.1 STRESS, STRAIN, AND MATERIAL PROPERTIES

Many of a material's structural properties are defined by their behavior under stress. A good understanding of these properties is needed to properly design for them.

Stress

If we apply an axial tensile or compressive force (P) on a material having a length (L) and a cross-sectional area (A) (Figure 3.1), the uniform *stress* (F) in the material in tension or compression is given by:

F = P/A

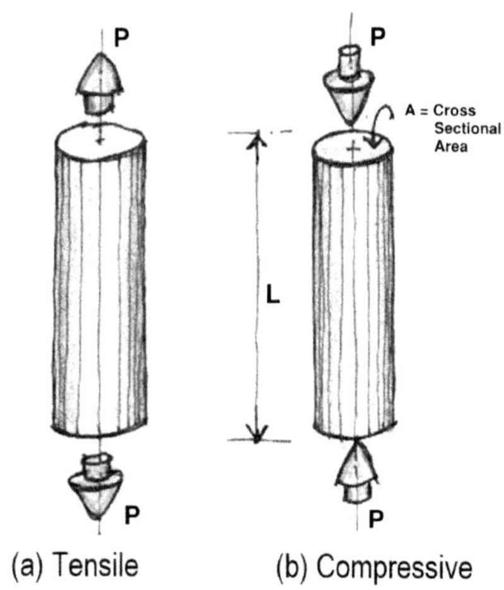

Figure 3.1 Stress in a Material

Strain

Under the applied load, the material will deform—i.e., it will elongate under tension, or shorten under compression (ΔL) (Figure 3.2). The amount of deformation divided by the original length is known as *strain* and is given by:

STRAIN = ΔL / L

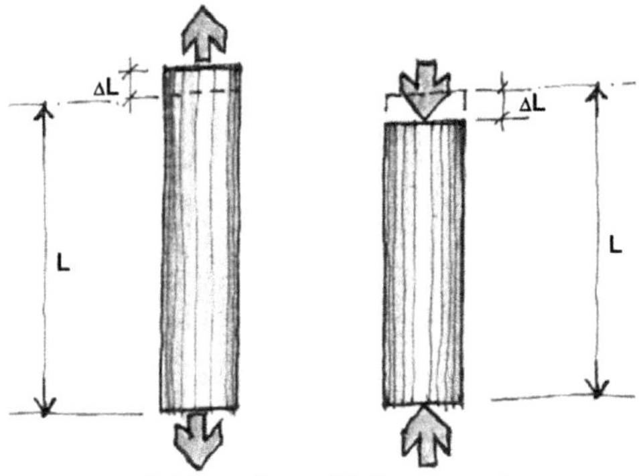

Figure 3.2 Strain in a Material

Stress-Strain Curves

A stress-strain curve is a plot of stress (along the vertical axis) vs. strain (along the horizontal axis) that helps to visualize important behavioral characteristics of a material under load (Figure 3.3). Every material has a unique stress-strain curve, in tension as well as in compression.

In a sense, a stress-strain curve can be considered a material's 'signature' behavior under load.

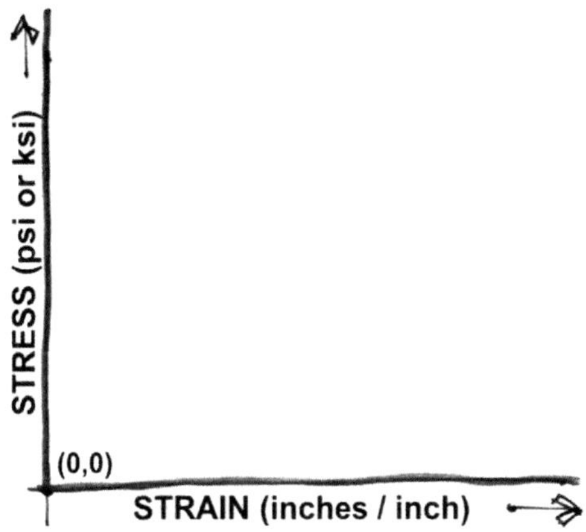

Figure 3.3 Axes of a Stress-Strain Curve

Material Strength

Material strength is the ability of a material to withstand applied loads. Loads on a member cause various internal stresses that tend to produce deformations in its material. Depending on the type of load (transverse, axial, or torsional), the member may experience compression, tension, shear, or a combination of these stresses.

- Steel's material strength is represented by various values, most importantly its yield stress (F_y).
- Wood's material strength is dependent upon several different criteria such as the direction of force, the direction of grain, and its species and grade. Wood's material strength is represented by values for bending stress (F_b), tensile strength parallel to the grain (F_t), compressive strength parallel to the grain (F_c), compressive strength perpendicular the grain ($F_{c\perp}$), and shear strength parallel to the grain (F_v).

- Concrete's most important material strength is its compressive strength, represented by (f'_c). Other important concrete properties such as shear strength, rupture strength, and modulus of elasticity, are expressed as a function of f'_c.

Material Stiffness

Robert Hooke in the late 1600s discovered that when stress is initially applied on a material, the strain increases proportionally. This proportional relationship, called Hooke's Law, is represented by a straight line on a stress-strain curve and exists only to the *proportional limit*—a unique point for different materials, beyond which they behave differently. Within the proportional limit, when stress is removed, strain disappears and the material returns to its original shape—the material is said to behave elastically.

Material stiffness is a measure of a material's resistance to deformation. It is represented by the *modulus of elasticity (E)*, also known as Young's modulus, and is commonly found in beam deflection and column buckling formulae. Modulus of elasticity is defined as the ratio of stress to strain *within the proportional limit* of a material's stress-strain curve, and is a linear relationship.

E = STRESS / STRAIN

The higher a material's modulus of elasticity, the steeper its stress-strain slope and the greater its stiffness (Figure 3.4).

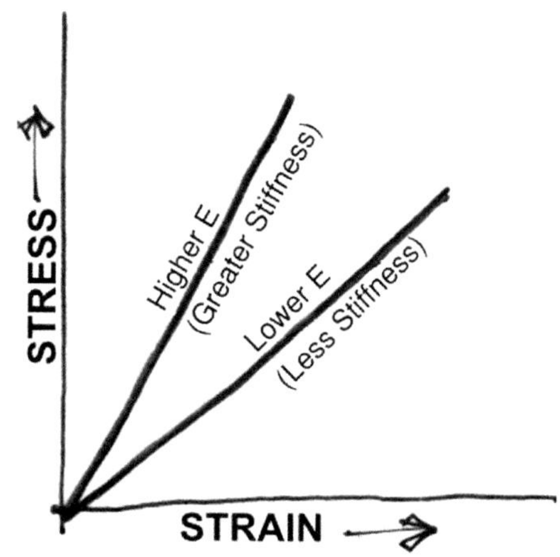

Figure 3.4 Modulus of Elasticity Comparison on a Stress-Strain Diagram

Let's examine the stress-strain curves for steel, wood, and concrete to better understand their unique structural characteristics.

3.2 STRESS-STRAIN CURVES FOR STEEL, WOOD, AND CONCRETE

Stress-Strain Curve for Steel in Tension

Steel's stress-strain curve is similar in tension and compression for stresses within the elastic range. Steel possesses *ductility*, a unique property that enables it to absorb large deformations beyond the elastic range. For this reason, steel's stress-strain curve in tension is the one of usual concern and the one we will look at in detail.

Steel's tensile stress-strain curve is based on laboratory tests of a cylindrical specimen pulled apart to failure.

An entire generalized curve for a typical steel specimen is shown in Figure 3.5.

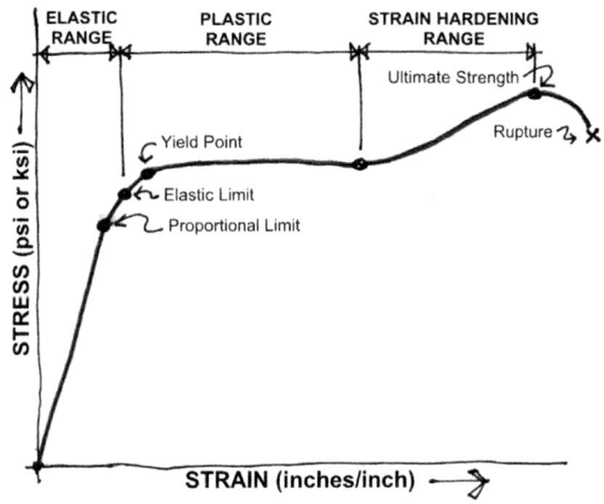

Figure 3.5 Steel's Stress-Strain Curve in Tension

Let's take a closer look.

When stress is initially applied, strain is directly proportional to stress, up to the proportional limit (Figure 3.5a).

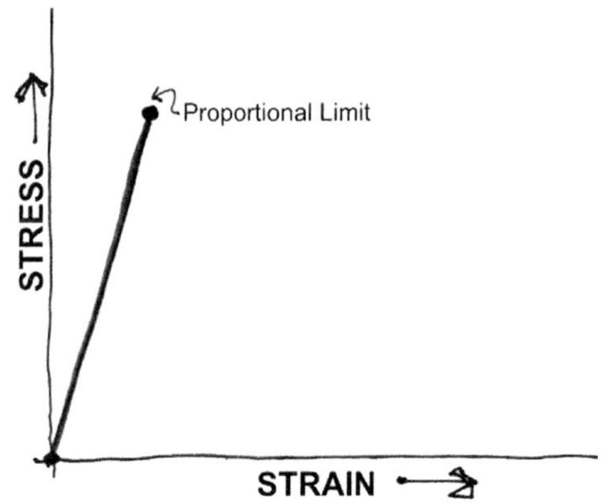

Figure 3.5a

Beyond the proportional limit, the relationship of stress to strain becomes non-linear. For every unit of stress, the increase in strain becomes greater. The upper limit of this behavior is termed the elastic limit.

Within the elastic limit, the strain disappears when the stress is removed, and the material returns to its original position without any permanent deformation (Figure 3.5b). Beyond the elastic limit, when the stress is removed, the material no longer returns to its original position and the material becomes permanently deformed. This deformation (i.e., strain) is termed *permanent set*.

For steel, the strain up to the elastic limit is termed the *elastic range* wherein the material is said to behave elastically.

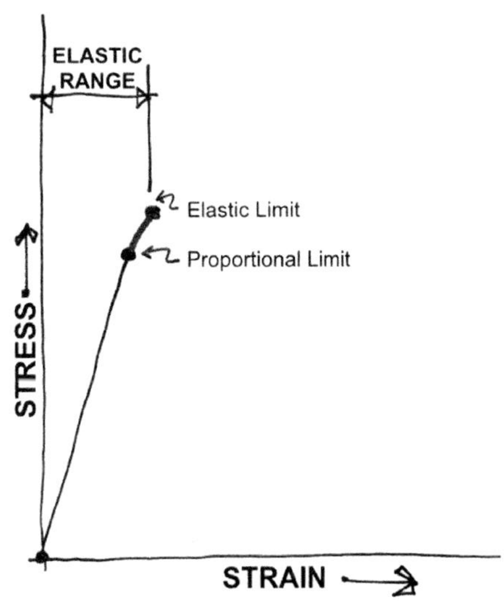

Figure 3.5b

With additional stress, the disproportionality of stress to strain continues to the yield point, the point at which the material is said to 'yield'—i.e., exhibit large increases in strain with little or no increase in stress (Figure 3.5c).

Note that steel's proportional limit, elastic limit, and yield point occur relatively close to each other, but they are distinct points for steel.

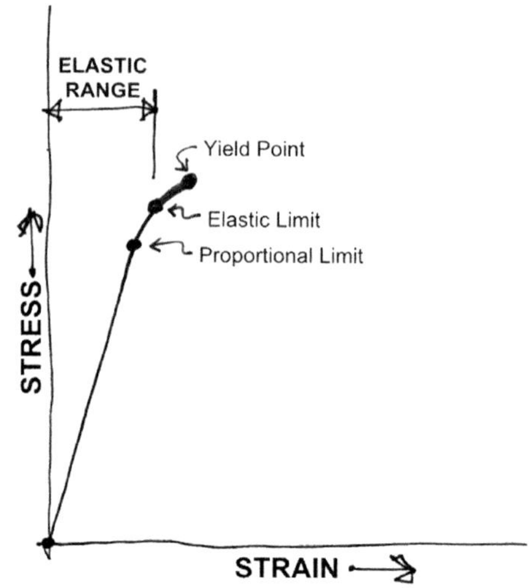

Figure 3.5c

Yielding continues throughout the plastic range until the start of strain hardening. For steel, the strain from the elastic limit to the start of strain hardening is termed the *plastic range*, wherein the material is said to behave plastically (Figure 3.5d).

Within this range, steel exhibits ductility.

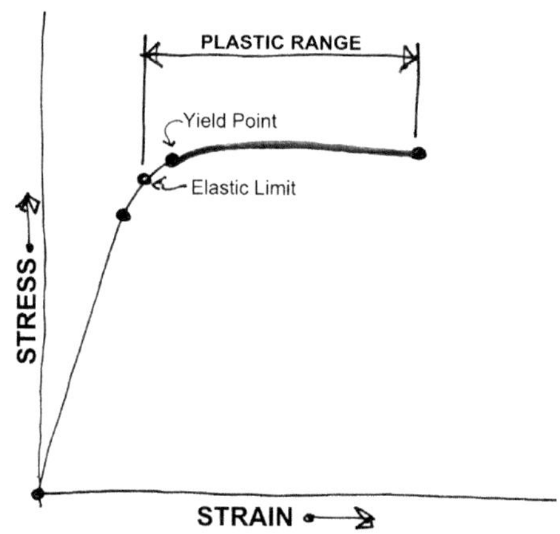

Figure 3.5d

As strain increases beyond the plastic range, the material enters the strain hardening range and is able to take additional stress with accompanying additional strain (Figure 3.5e).

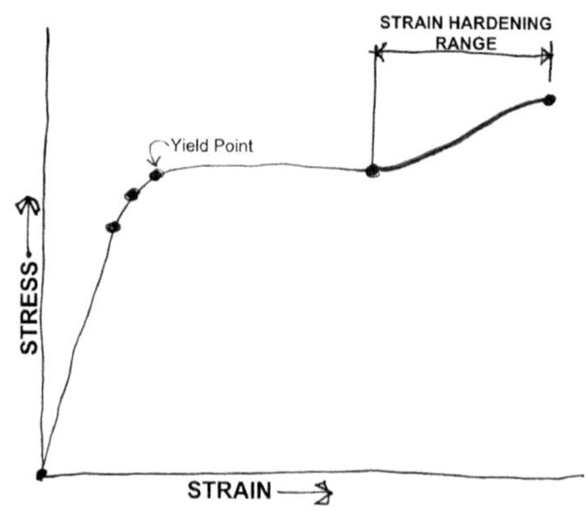

Figure 3.5e

The upper limit of the strain hardening range is the material's ultimate strength, beyond which the material begins to 'neck' until rupture (Figure 3.5f).

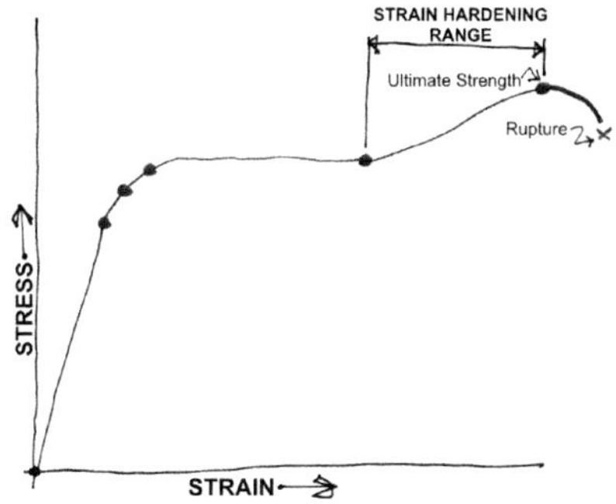

Figure 3.5f

Necking, a phenomenon exhibited by ductile materials, is characterized by a localized stretching with a corresponding reduction in the cross-sectional area. Once begun, with no increase in stress, necking continues until rupture (Figure 3.6).

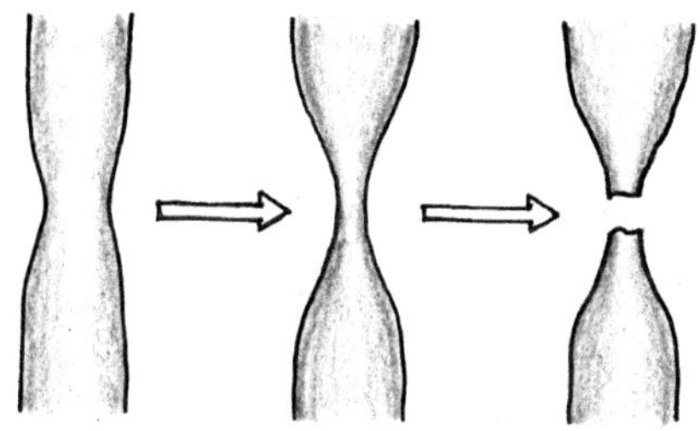

Figure 3.6 Necking Leading to Rupture

Stress-Strain Curve for Wood in Compression

Wood's properties vary according to species, grade, and direction of grain. Its non-homogenous nature makes its stress-strain curve in tension somewhat less predictable. For this reason, wood's stress-strain curve in compression is the one of usual concern and the one we'll examine.

Wood's compressive stress-strain curve is based on laboratory tests of a specimen compressed to failure. An entire generalized curve for a typical wood specimen is shown in Figure 3.7.

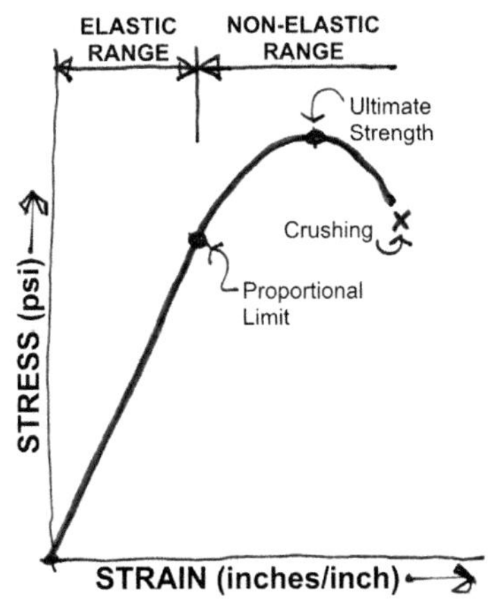

Figure 3.7 Wood's Stress-Strain Curve in Compression

Let's take a closer look.

When stress is initially applied, strain is directly proportional to stress up to the proportional limit (Figure 3.7a).

For wood, the strain up to the proportional limit is considered the elastic range—the strain disappears when the stress is removed. Wood's proportional limit is approximately 75% of its ultimate strength.

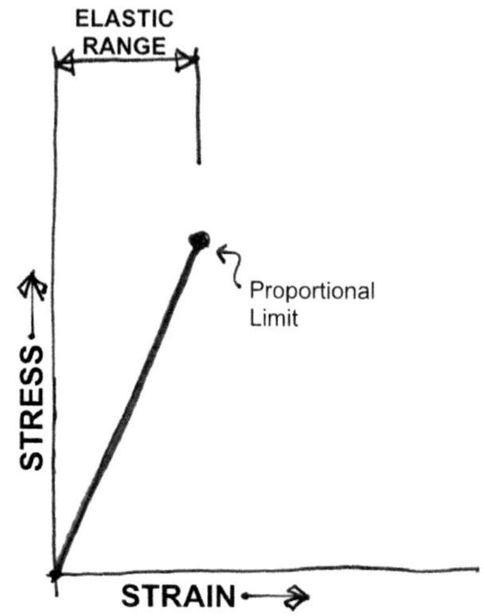

Figure 3.7a

Beyond the proportional limit, the relationship of stress to strain becomes non-linear. For every unit of stress, the increase in strain becomes greater. This behavior continues to the material's ultimate strength, at which point the material continues to strain with little or no additional stress until crushing.

The strain beyond the proportional limit is considered the non-elastic range for wood—the strain remains when the stress is removed (Figure 3.7b).

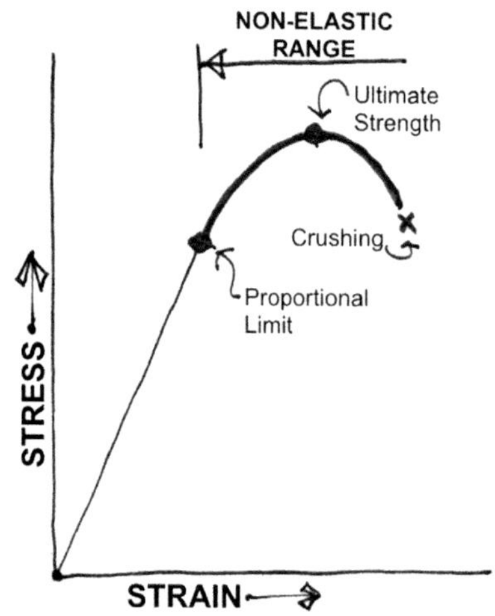

Figure 3.7b

Stress-Strain Curve for Concrete in Compression

Concrete is strong in compression and weak in tension. Its minimal tensile strength is typically ignored in structural applications. Concrete is also brittle, a property that causes it to fracture suddenly without warning when overstressed. For these reasons, concrete's stress-strain curve in compression is the one of usual concern and the one we'll examine.

Concrete's compressive stress-strain curve is based on laboratory tests of a specimen compressed to failure. An entire generalized stress-strain curve for a typical concrete specimen is shown in Figure 3.8.

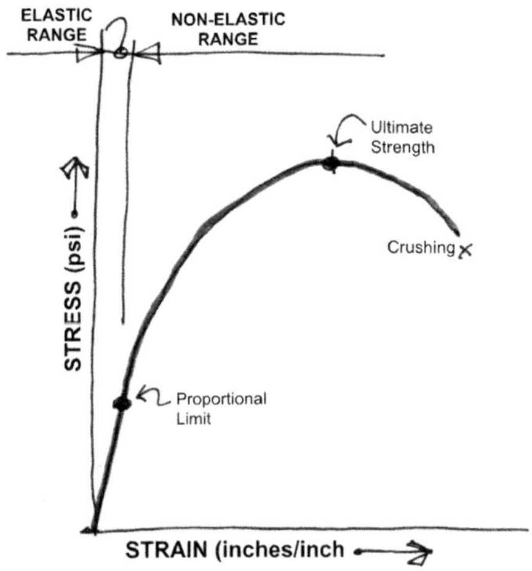

Figure 3.8 Concrete's Stress-Strain Curve in Compression

Let's take a closer look.

When stress is initially applied, strain is directly proportional to stress up to the proportional limit (Figure 3.8a).

For concrete, the strain up to the proportional limit is considered the elastic range—the strain disappears when the stress is removed. Concrete's proportional limit is approximately 30–40% of its ultimate strength.

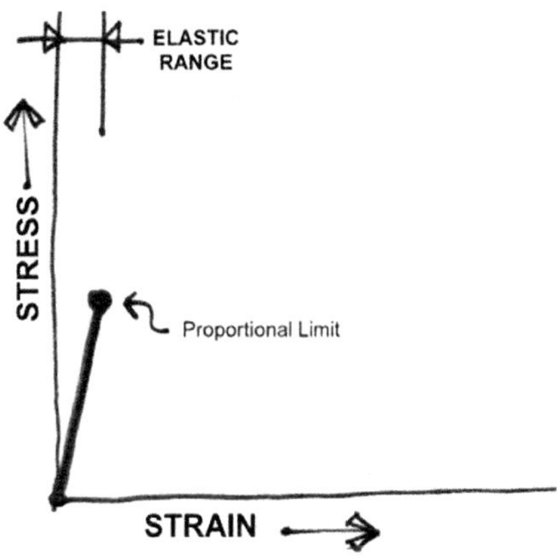

Figure 3.8a

Beyond the proportional limit, the relationship of stress to strain becomes non-linear. For every unit of stress, the increase in strain becomes greater. This behavior continues to the material's ultimate strength, at which point the material continues to strain with no additional stress until crushing (Figure 3.8b).

Within the range of 3,000 to 6,000 psi concrete, concrete's ultimate strength is achieved at a strain of approximately 0.002, and crushes at a strain between 0.003 and 0.004.

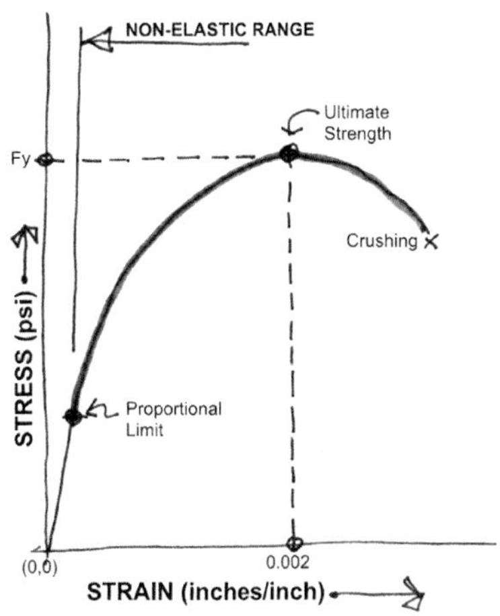

Figure 3.8b

Comparison of Stress-Strain Curves for Steel, Wood, and Concrete

As a generalized comparison, Figure 3.9 shows the approximate stress-strain curves for steel in tension, wood in compression, and concrete in compression.

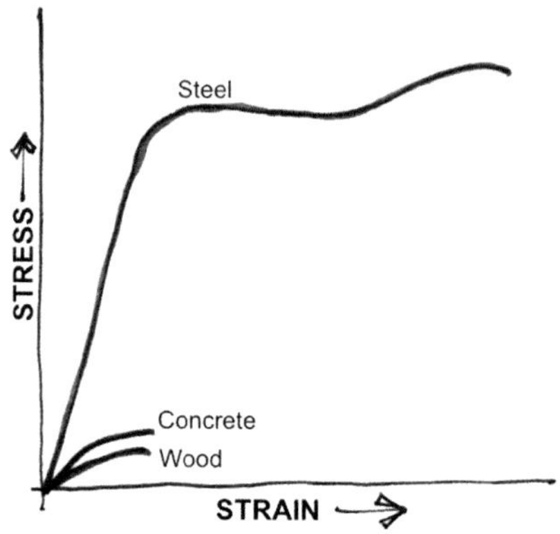

Figure 3.9 Comparative Stress-Strain Curves for Steel, Wood, Concrete

3.3 ELASTIC/PLASTIC MATERIAL BEHAVIOR—AN ANALOGY

To get a better intuitive sense for elastic and plastic material behavior, let's begin by envisioning a group of workers holding a large, flat panel above their heads (Figure 3.10).

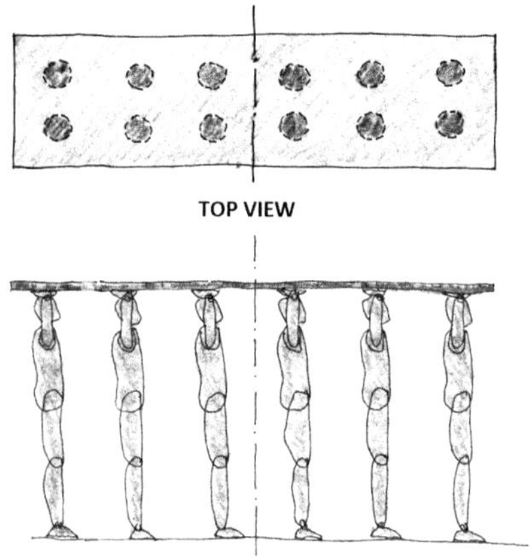

Figure 3.10 Workers Holding a Panel

Situation 1—Elastic Behavior

If a wind gust occurs, the panel could push down on the workers on the left side and pull up on the workers on the right side (Figure 3.11).

The workers on the left would be compressed (pushing the panel up), while the workers on the right would be stretched (holding the panel down). The workers on the far left are compressed the most, while the workers on the far right are stretched the most.

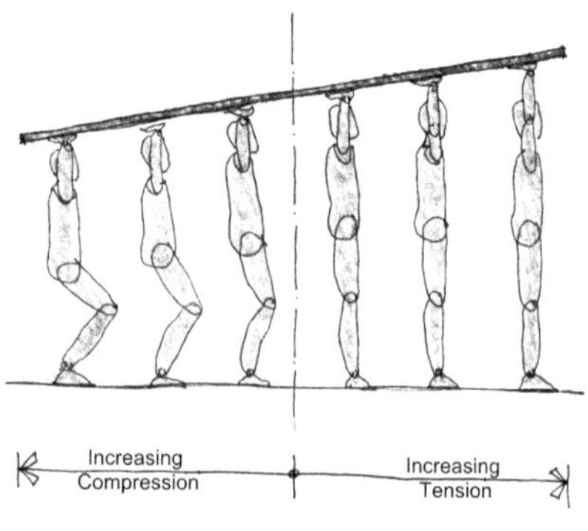

Figure 3.11 Workers under Stress

The forces acting on the workers are diagrammatically shown in Figure 3.12.

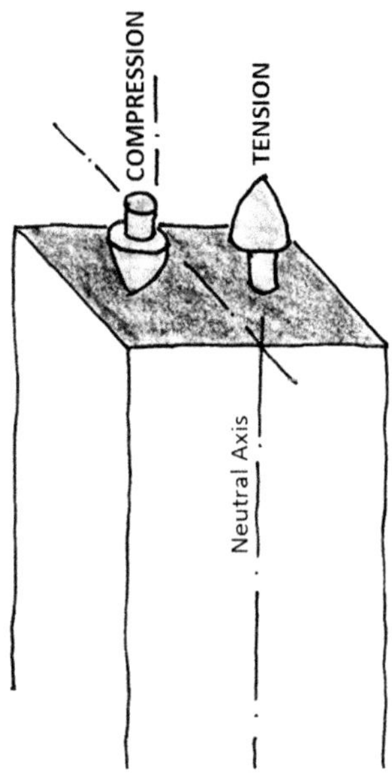

Figure 3.12 Compressive and Tensile Forces on the Workers

The stress distribution diagram of the workers is shown in Figure 3.13. The workers on the far left and far right respectively experience the greatest compressive and tensile stress, gradually diminishing from worker to worker towards the center.

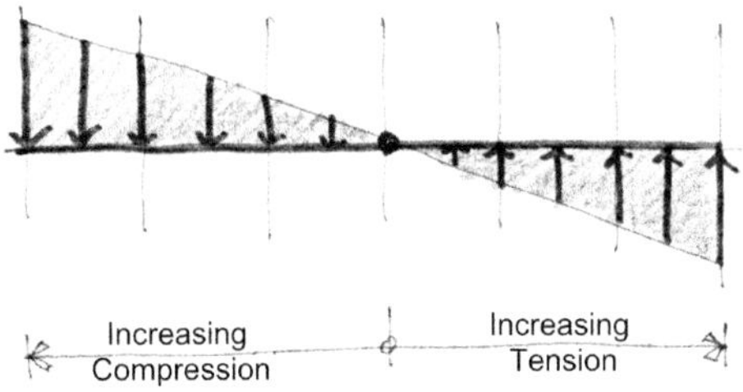

Figure 3.13 Stress Distribution Diagram

Situation 2—Plastic Behavior

If the wind gust gradually increases in intensity thereby exerting additional pressure, the workers at the extreme ends will eventually reach the limit of their respective compressive and tensile strength. At this point, the workers just inside them will begin to support the additional load (Figure 3.14).

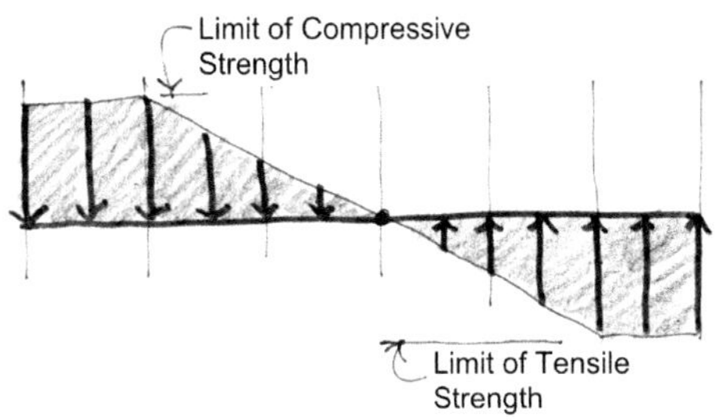

Figure 3.14

This process continues towards the center (Figure 3.15).

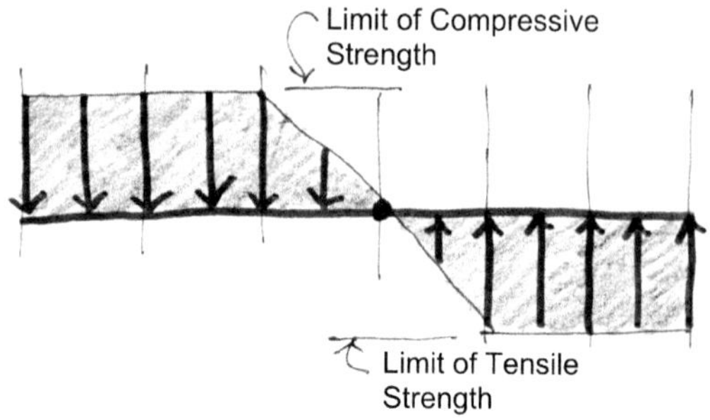

Figure 3.15

Eventually all the workers become fully and equally stressed to the limit of their strength (Figure 3.16).

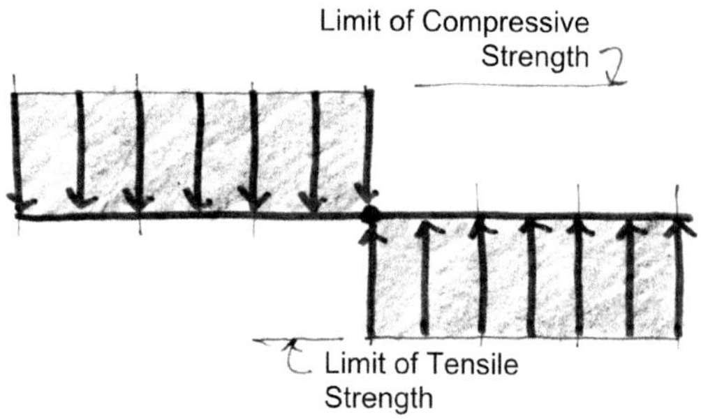

Figure 3.16

Discussion

In Situation 1, only the workers at the extreme ends are stressed to the limit of their strength—in a sense their yield strength. The group has limited their collective load-carrying capacity to the strength of the workers at the extreme ends. We can say we will not stress the group any higher, but this leaves a lot of reserve strength not being used (i.e., the strength of the workers on the inside). In Situation 1 the workers as a group can be considered stressed to their elastic limit.

In Situation 2, all workers are fully stressed to the limit of their strength. The group has maximized their collective load-carrying capacity by equally contributing to the full limit of their individual strengths. We cannot stress the workers any higher without failure, since there is no reserve strength. In Situation 2 the workers as a group can be considered stressed to their ultimate strength.

For a proper margin of safety, it's apparent that greater safety factors would be required in Situation 2 than Situation 1.

3.4 ELASTIC/PLASTIC ANALOGY APPLIED TO A BEAM

Let's correlate the analogy of the group of workers to a simple steel beam subject to bending under load (Figure 3.17).

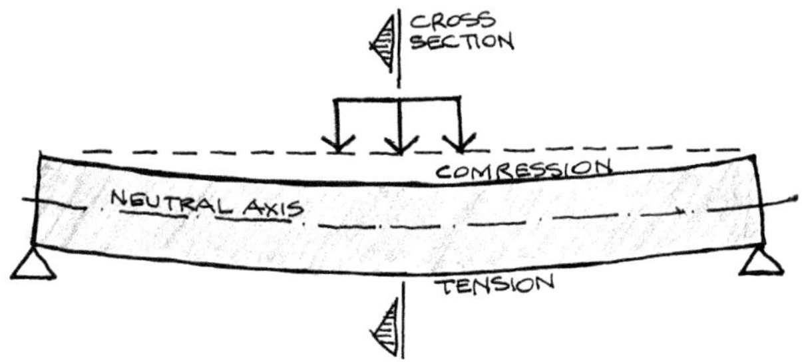

Figure 3.17 Simple Beam Bending Under Load

The beam's cross section throughout the member is subject to compression above the neutral axis and tension below the neutral axis (Figure 3.18)—with the greatest compressive and tensile stresses occurring at the *beam's* midspan, where the bending moment is the maximum.

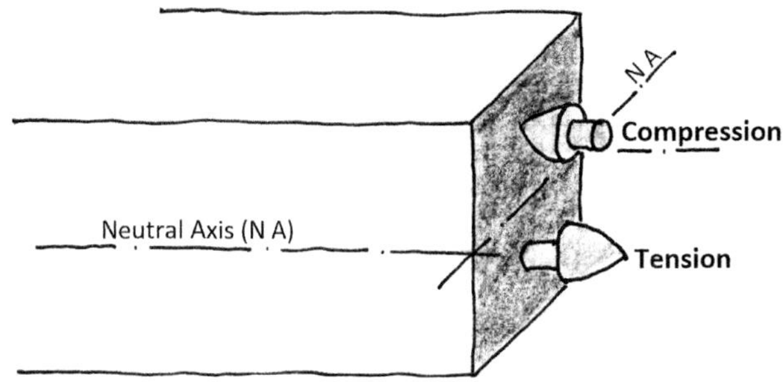

Figure 3.18 Cross Section

Situation 1—Elastic Behavior

With the beam stressed to its elastic limit, the beam section's stress distribution diagram is as shown in Figure 3.19. The extreme fibers at the top and bottom of the section respectively take the greatest compressive and tensile stress, proportionally diminishing to zero at the neutral axis.

At the elastic limit, the extreme fibers are stressed to their yield point (F_y).

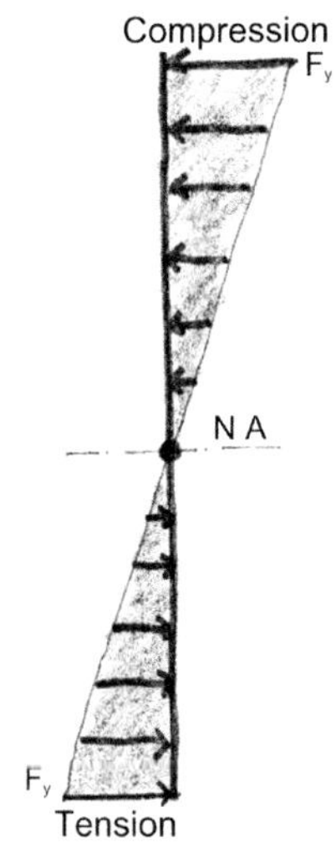

Figure 3.19

Situation 2—Plastic Behavior

With the beam stressed beyond the elastic limit, the extreme fibers cannot take any more stress, and the interior fibers begin to carry additional load until they too reach their yield point (Figure 3.20).

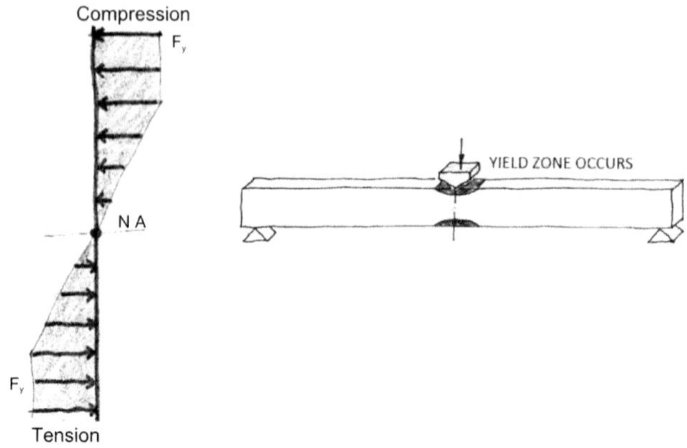

Figure 3.20 Initial Plastic

This process continues towards the neutral axis (Figure 3.21).

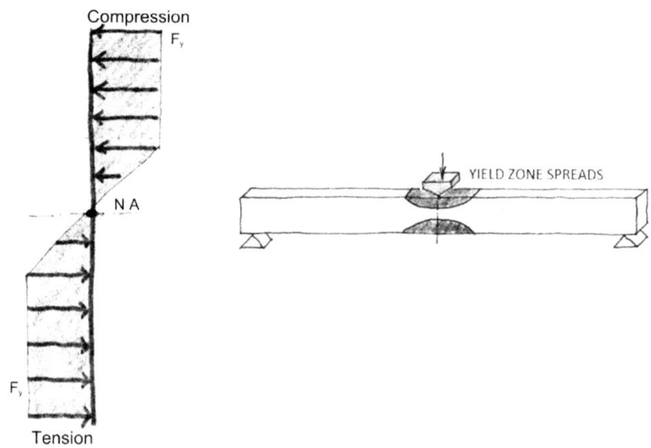

Figure 3.21 Initial Plastic

Eventually all fibers become fully and equally stressed to their yield point (Figure 3.22).

When this occurs, the beam is said to have reached its plastic limit (i.e., the limit of material resistance). The beam yields by forming a 'plastic hinge', at the point of maximum moment.

The maximum moment that the beam can resist at the plastic limit is termed the *plastic moment*.

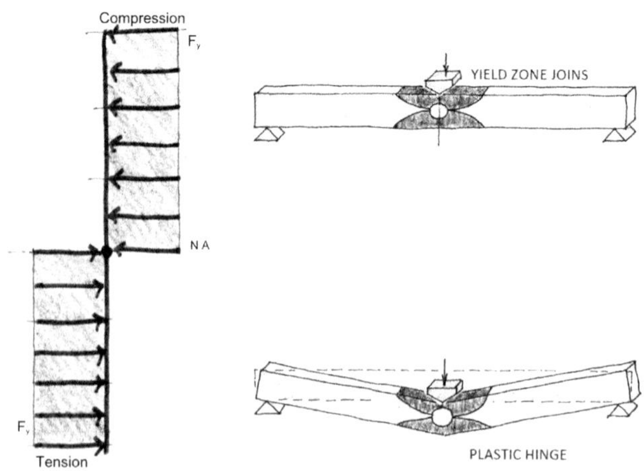

Figure 3.22 Fully Plastic

Materials and Elastic/Plastic Design

In Allowable Stress Design, strains are kept within the elastic range with stresses at the extreme fibers of the cross section not exceeding their yield point. This design process is termed *elastic design*.

In Allowable Strength Design (ASD) and in Load and Resistance Factor Design (LRFD), design strains are taken beyond the elastic range with stresses throughout the cross section being at the yield point. This design process is termed *plastic design*.

In Strength Design, the interaction of concrete and steel creates a more complex situation wherein the steel is stressed to its yield stress, and the concrete is strained to code-specified limits. This will be discussed in greater detail in Chapter 10.

Regardless of the design methodology or material theory, keep in mind that appropriate safety factors must be applied to assure that the actual stresses and strains experienced by a member fall safely within the elastic range.

3.5 SECTION MODULUS

Section modulus, measured by inches to the third power (in^3), is a property of a cross section that is used in design formulae for flexural members such as beams. There are two kinds of section moduli, elastic and plastic.

Elastic Section Modulus (S)

The elastic section modulus S is used in conjunction with elastic design. It is based on the idea that resistance to bending is limited by the yielding of the extreme fibers in a member's cross section—in other words the fibers farthest from the neutral axis. It is given by the formula:

S = I / c where:

S = elastic section modulus
I = moment of inertia
c = distance of farthest fiber from neutral axis (Figure 3.23)

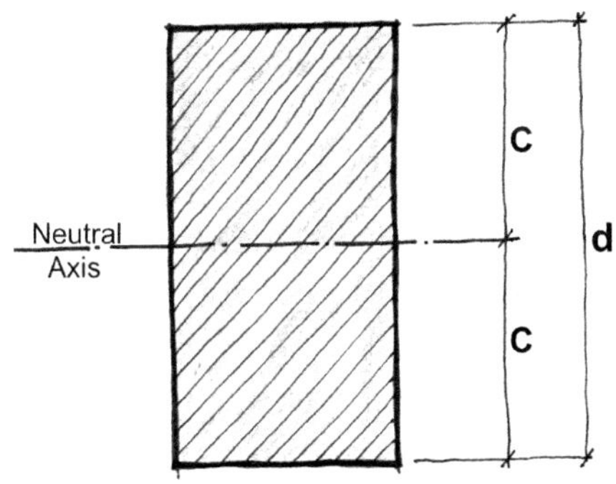

Figure 3.23 "c" in a Rectangular Beam

Plastic Section Modulus (Z)

The plastic section modulus Z is used in conjunction with plastic design. It is based on the idea that resistance to bending is limited by the yielding of all the fibers in a material's cross section—half in compression on one side of the neutral axis and half in tension on the other side.

Elastic and Plastic Neutral Axis

The neutral axis of a section is not necessarily coincident under elastic and plastic conditions.

In a cross section symmetrical about its xx neutral axis such as a rectangle, the elastic and plastic neutral axes are the same, with elastic and plastic stress distributions as shown in Figure 3.24.

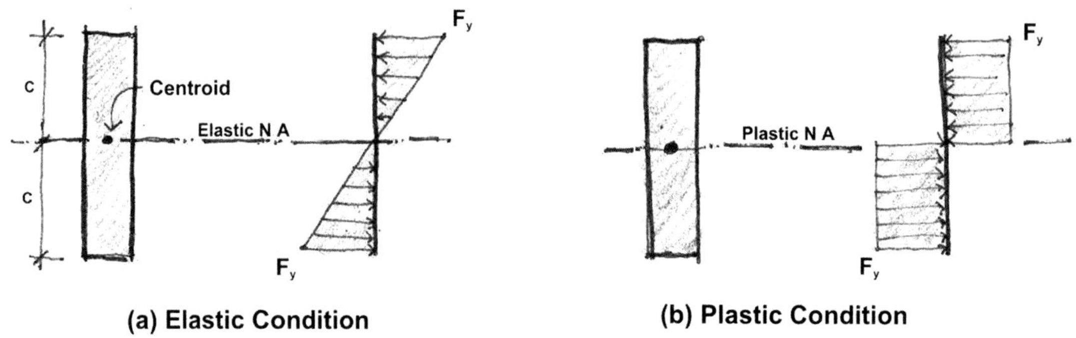

Figure 3.24 The Neutral Axis of a Symmetrical Section

In a cross section asymmetrical about its xx neutral axis such as a tee, the elastic neutral axis can be defined as a line passing through the section's centroid, while the plastic neutral axis can be defined as a line that divides the area of the cross section into two equal parts such that the total compressive force C is equal to the total tensile force T. Stress distributions for these conditions are as shown in Figure 3.25.

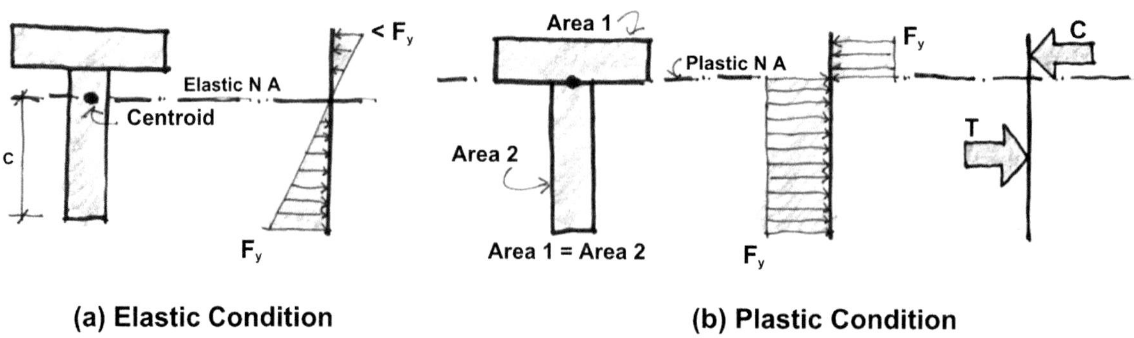

Figure 3.25 The Neutral Axis of an Asymmetrical Section

Shape Factor

A section's shape factor is defined as the ratio of Z to S. The shape factor is an indication of the comparative moment capacity of a given section in plastic vs. elastic design. For a rectangular section the shape factor = 1.5, indicating that the plastic moment capacity is 1.5 times the elastic moment capacity.

3.6 GETTING STARTED—CASE STUDY

The best way to see the practical application of the various design methodologies is by example. For this purpose we'll use Case Studies of diagrammatic two-story structures framed in steel, wood, and reinforced concrete, and design typical structural members. We'll start by examining the unique characteristics of each material, and then see their application in each of the Case Studies. But before we start, let's review a few overriding structural concepts.

Loads, Stress, and Resistance

Loads on a member create tensile, compressive, and shear stresses within it—these stresses are resisted by the properties of the member's material and cross section. For the design of a member, the resistance to stress provided by its material and cross section must be equal to or greater than the stress created by the loads (Figure 3.26).

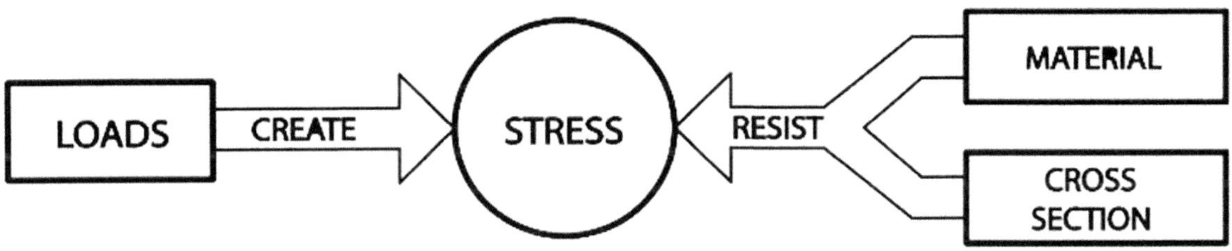

Figure 3.26 Loads, Stress, Material, and Cross Section

Fundamental Formulae

Two fundamental structural formulae relate material resistance, cross-sectional resistance, and stress.

For Compression and Tension Members

F = P / A where:

F = stress (i.e., material resistance)
P = a direct compressive or tensile force (i.e., a load)
A = cross-sectional area (i.e., cross-sectional resistance)

For Flexural Members

F = M / S	where:
(for elastic design)	F = stress (i.e., material resistance)
or	M = moment (created by a load)
F = M / Z	S = elastic section modulus (i.e., cross-sectional resistance)
(for plastic design)	Z = plastic section modulus (i.e., cross-sectional resistance)

Another fundamental formula concerns moment, either external or internal, on a member:

MOMENT = FORCE × DISTANCE

As we go through each Case Study, we'll frequently see variations of these formulae in use. Keep the simplicity of these formulae and their underlying concepts in mind as we examine the various factors that govern the structural design of a member.

Let's get started.

4 Case Study Introduction

Our Case Studies will be a two-story building with one-way decks over beams and girders, framing into columns. We'll focus on the first floor framing and design a typical beam, girder, and column in the three common structural materials—steel, wood, and concrete. Floor loads will be applied onto a beam, based upon the beam's load tributary area. The beam's reactions will then be determined and applied onto girders. Beam and girder reactions, along with roof loads from above, will then be applied onto an interior column below.

An adaptation of the diagrammatic structure in Figure 4.1 will be used in each Case Study. The purpose of the diagram is to aid in presenting design principles and should not be taken as a literal representation of a building.

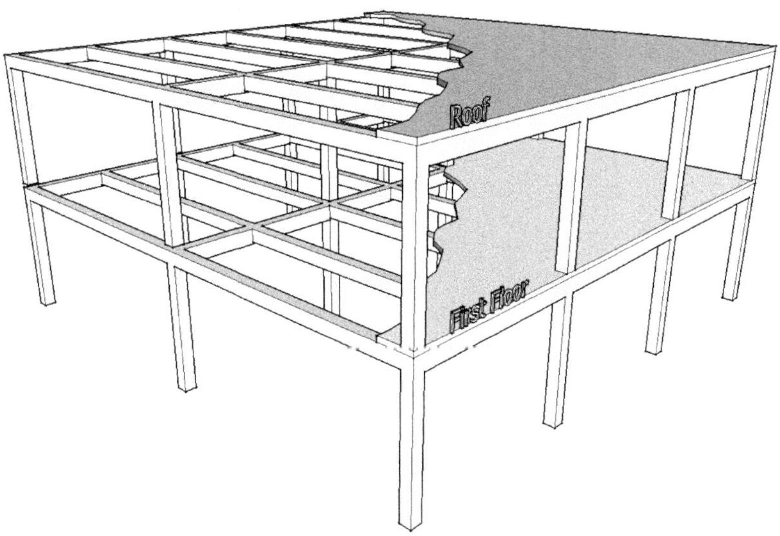

Figure 4.1 Diagrammatic Structure for Case Studies

4.1 LOAD COMBINATIONS

Recall from Chapter 2 that load combinations are prescribed according to the design methodology being used, and that the load combinations resulting in the most critical effect are used to compute the forces and moments acting on a structure. The service loads used in load combinations are:

D = dead load	L_r = roof live load	R = rain, ponding, ice load	E = earthquake load
L = floor live load	S = snow load	W = wind load	

Our Case Studies will consider vertical dead loads, and vertical floor and roof live loads only (D, L, and L_r respectively). For simplicity, we will also not consider lateral loads from wind and earthquake. We'll therefore use zero as the value for S, R, W, and E in the load combinations.

IBC-Prescribed Load Combinations

Tables 4.1 and 4.2 show IBC-prescribed load combinations. Table 4.1 is for use in both Allowable Strength Design (ASD) and Allowable Stress Design. Table 4.2 is for use in both Load and Resistance Factor Design (LRFD) and Strength Design. These tables also show, shaded, the derivation of the governing load combinations for the Case Studies according to our assumptions.

General Design Approach for Structural Members

In essence, the design of a structural member in a given material involves the selection of a cross-sectional shape and size that is adequate for the loads imposed upon it. The member should not be over-designed (i.e., should not have significantly more strength than required), and it should definitely not be under-designed (i.e., have less strength than necessary). Over-designing is uneconomical and under-designing is unsafe.

Table 4.1 IBC-Prescribed Load Combinations

For Use with Allowable Strength Design (ASD), and Allowable Stress Design

	Load Combinations Applied to Floors		
Load Combinations	**Substituting Zero for Lr, S, R, W, E**	**Load Combination Reduces to**	**Case Study Governing Load Combination**
D	D	D	
D + L	D + L	D + L	D + L
D + (L_r or S or R)	D + ~~(L_r or S or R)~~	D	
D + 0.75 L + 0.75 (L_r or S or R)	D + 0.75 L + ~~0.75 (L_r or S or R)~~	D + 0.75 L	
D + (0.6 W or 0.7 E)	D + ~~(0.6 W or 0.7 E)~~	D	
D + 0.75 L + 0.75 (0.6W) + 0.75 (L_r or S or R)	D + 0.75 L + ~~0.75 (0.6W)~~ + ~~0.75 (L_r or S or R)~~	D + 0.75 L	
D + 0.75 L + 0.75 (0.7 E) + 0.75 S	D + 0.75 L + ~~0.75 (0.7 E)~~ + ~~0.75 S~~	D + 0.75 L	
0.6 D + 0.6 W	0.6 D + ~~0.6 W~~	0.6 D	
0.6 D + 0.7 E	0.6 D + ~~0.7 E~~	0.6 D	

	Load Combinations Applied to Roofs		
Load Combinations	**Substituting Zero for L, S, R, W, E**	**Load Combination Reduces to**	**Case Study Governing Load Combination**
D	D	D	
D + L	D + ~~L~~	D	
D + (L_r or S or R)	D + (L_r ~~or S or R~~)	D + L_r	D + L_r
D + 0.75 L + 0.75 (L_r or S or R)	D + ~~0.75 L~~ + 0.75 (L_r ~~or S or R~~)	D + 0.75 L_r	
D + (0.6 W or 0.7 E)	D + ~~(0.6 W or 0.7 E)~~	D	
D + 0.75 L + 0.75 (0.6W) + 0.75 (L_r or S or R)	D + ~~0.75 L~~ + ~~0.75 (0.6W)~~ + 0.75 (L_r ~~or S or R~~)	D + 0. 75 L_r	
D + 0.75 L + 0.75 (0.7 E) + 0.75 S	D + ~~0.75 L~~ + ~~0.75 (0.7 E)~~ + ~~0.75 S~~	D	
0.6 D + 0.6 W	0.6 D + ~~0.6 W~~	0.6 D	
0.6 D + 0.7 E	0.6 D + ~~0.7 E~~	0.6 D	

Table 4.2 IBC-Prescribed Load Combinations

For Use with Load and Resistance Factor Design (LRFD), and Strength Design

	Load Combinations Applied to Floors		
Load Combinations	**Substituting Zero for L_r, S, R, W, E**	**Load Combination Reduces to**	**Case Study Governing Load Combination**
1.4 D	1.4 D	1.4 D	
1.2 D + 1.6 L + 0.5 (L_r or S or R)	1.2 D + 1.6 L + ~~0.5 (L_r or S or R~~)	1.2 D + 1.6 L	1.2 D + 1.6 L
1.2 D + 1.6 (L_r or S or R) + (0.5 L or 0.5 W)	1.2 D + ~~1.6 (L_r or S or R~~) + (0.5 L ~~or 0.5 W~~)	1.2 D + 0.5 L	
1.2 D + 1.0 W + 0.5 L + 0.5 (L_r or S or R)	1.2 D + ~~1.0 W~~ + 0.5 L + ~~0.5 (L_r or S or R~~)	1.2 D + 0.5 L	
1.2 D + 1.0 E + 0.5 L + 0.2 S	1.2 D + ~~1.0 E~~ + 0.5 L + ~~0.2 S~~	1.2 D + 0.5 L	
0.9 D + 1.0 W	0.9 D + ~~1.0 W~~	0.9 D	
0.9 D + 1.0 E	0.9 D + ~~1.0 E~~	0.9 D	

	Load Combinations Applied to Roofs		
Load Combinations	**Substituting Zero for L, S, R, W, E**	**Load Combination Reduces to**	**Case Study Governing Load Combination**
1.4 D	1.4 D	1.4 D	
1.2 D + 1.6 L + 0.5 (L_r or S or R)	1.2 D + ~~1.6 L~~ + 0.5 (L_r ~~or S or R~~)	1.2 D + 0.5 L_r	
1.2 D + 1.6 (L_r or S or R) + (0.5 L or 0.5 W)	1.2 D + 1.6 (L_r ~~or S or R~~) + (~~0.5 L or 0.5 W~~)	1.2 D + 1.6 L_r	1.2 D + 1.6 L_r
1.2 D + 1.0 W + 0.5 L + 0.5 (L_r or S or R)	1.2 D + ~~1.0 W + 0.5 L~~ + 0.5 (L_r ~~or S or R~~)	1.2 D + 0.5 L_r	
1.2 D + 1.0 E + 0.5 L + 0.2 S	1.2 D + ~~1.0 E + 0.5 L + 0.2 S~~	1.2 D	
0.9 D + 1.0 W	0.9 D + ~~1.0 W~~	0.9 D	
0.9 D + 1.0 E	0.9 D + ~~1.0 E~~	0.9 D	

4.2 DESIGNING FOR BEAMS AND GIRDERS

A beam or girder is typically a horizontal member primarily subject to bending (flexure). Its design is dependent upon its material, cross-sectional properties, loads, span, and supports.

Deeper beams tend to be more economical in terms of weight and, in the absence of other controlling considerations, are typically selected. However other factors for appropriate beam depth must also be considered such as controlling floor-to-floor and total building heights.

Case Study Assumptions

For simplicity, beams and girders will be considered to have:

- simple supports
- full lateral bracing

General Design Approach

The general design approach for a beam or girder involves confirming its adequacy for *strength* (primarily moment and shear capacity) and *serviceability* (primarily deflection). The typical steps are:

1. designing to resist moment
2. checking for shear in steel and wood; designing for shear in reinforced concrete
3. checking for deflection

1. Designing to Resist Moment

The moment in a beam is the internal stress produced by loads causing it to bend (Figure 4.2).

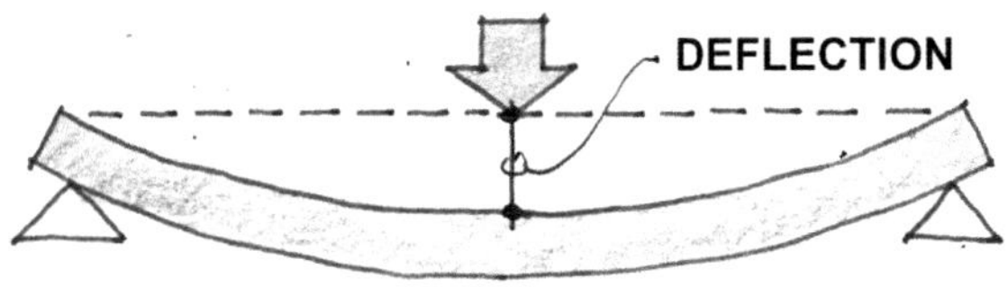

Figure 4.2 Bending and Deflection in a Simple Beam

Determining a beam's required cross section to resist moment is typically the starting point in its design. This involves determining a suitable cross section with an available flexural capacity equal to or greater than that required.

- For steel and wood, we'll calculate the required section modulus and then select a cross-sectional shape with an available section modulus equal to or greater than that required. The AISC and NDS both provide beam selection tables that facilitate the selection of beam sizes.
- For concrete, we'll assume a suitable cross-sectional shape for the beam and then confirm that its available flexural capacity is equal to or greater than that required.

2. Checking/Designing for Shear

The shear in a beam is the internal stress produced by the loads causing it to become sliced. Shear stresses can be vertical or horizontal (Figures 4.3a and 4.3b).

Vertical shear is usually the main concern in steel and concrete beams, while horizontal shear is usually of greater concern in wood beams.

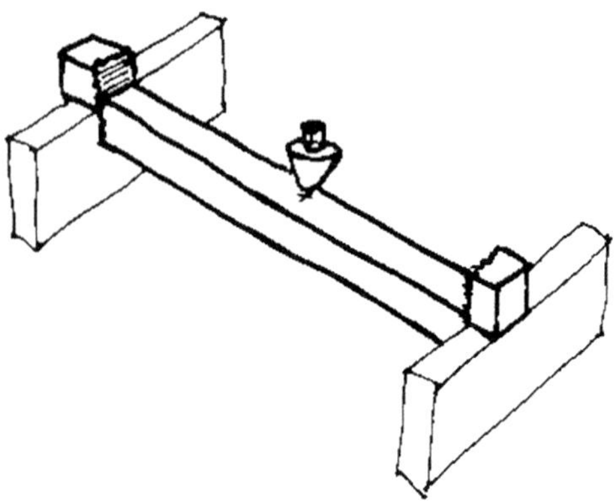

Figure 4.3a Vertical Shear in a Simple Beam

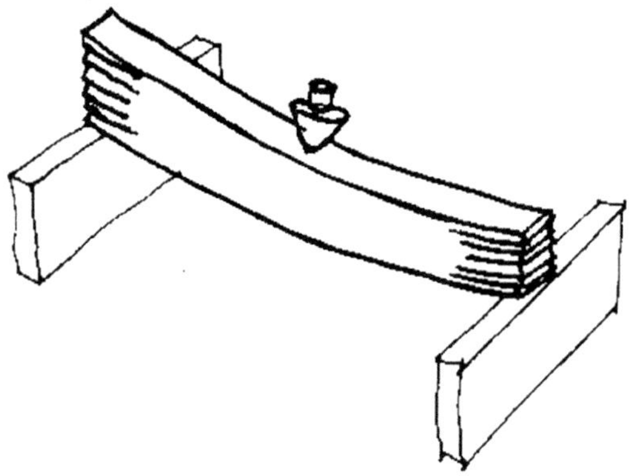

Figure 4.3b Horizontal Shear in a Simple Beam

Shear in Steel and Wood

Resisting maximum moment in a typical steel or wood beam of moderate span often results in a section of adequate strength to resist shear. Nevertheless, shear must still be checked. Checking for shear in a steel or wood beam involves assuring that its available shear strength is equal to or greater than that required.

Shear in Concrete

Shear in a reinforced concrete beam is somewhat more complex since it involves the interaction of two materials. Steel reinforcing (stirrups) is used to supplement the shear capacity of the concrete. We'll analyze how these two materials interact in Chapters 10 and 11.

Shear is of greater concern when large concentrated loads occur near the supports, or when a large concentrated load occurs over a short span (Figure 4.4).

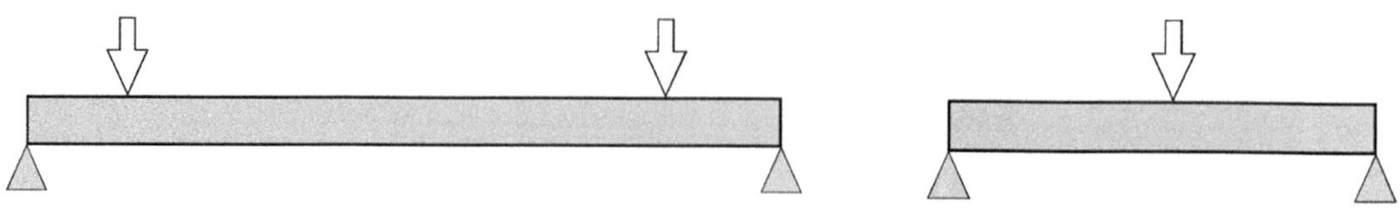

(a) Large Concentrated Loads Near the Supports

(b) A Large Concentrated Load Over a Short Span

Figure 4.4 Conditions of Greater Concern for Shear in a Beam

3. *Checking for Deflection*

Deflection in a beam is the vertical displacement of a point on the member due to an applied load (Figures 4.2 and 4.5).

Figure 4.5 Laboratory Test of Concrete Beam Exhibiting Deflection

Resisting maximum moment in a typical beam of moderate span generally results in a section adequate to limit excessive deflection. Nevertheless, deflection must still be checked.

Excessive deflection can have numerous harmful effects on a structure such as sagging floors, ponding on roofs, and cracks in ceilings and walls. Excessive vibration is generally a concern where mechanical equipment is supported over long spans not having benefit of walls or other vibration damping systems. For our Case Studies, since we'll be dealing with vertical loads only, we'll simply check for vertical deflection and not consider the effects of lateral deflection and vibration.

Checking for deflection involves assuring that the beam's *maximum deflection is equal to or less than its allowable deflection* for both (Figure 4.6):

- live load
- total load (dead + live)

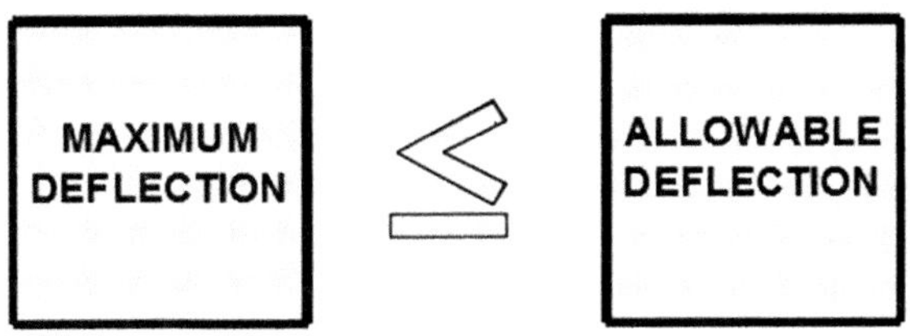

Figure 4.6 Maximum and Allowable Deflection

Maximum and Allowable Deflection

The *maximum* deflection of any beam is given by formulae that are dependent upon the beam's span, material, cross section, loading, and support conditions. Service loads are used to determine maximum deflection in all design methodologies.

The allowable deflection of any beam is limited by building codes to formulae that are dependent upon its span, whether it is a floor or roof member, and the type of finishes it supports. For the floor beams and girders in the steel and wood Case Studies, we'll be working with the deflection formulae and allowable deflection limits in Figures 4.7 and 4.8. Deflection in reinforced concrete beams will be addressed in Chapters 10 and 11.

Maximum Deflection Formulae

Figure 4.7 shows the formulae for maximum deflection at midspan, of the simply supported beams and girders we'll encounter in our Case Studies.

Maximum deflection under a uniform load (Figure 4.7a):

$$\boldsymbol{\delta = 5(wL^4) / 384\ EI}$$

For convenience in calculations, this can be expressed as:

$$\boldsymbol{\delta = 5(wL \times L^3) / 384\ EI}$$

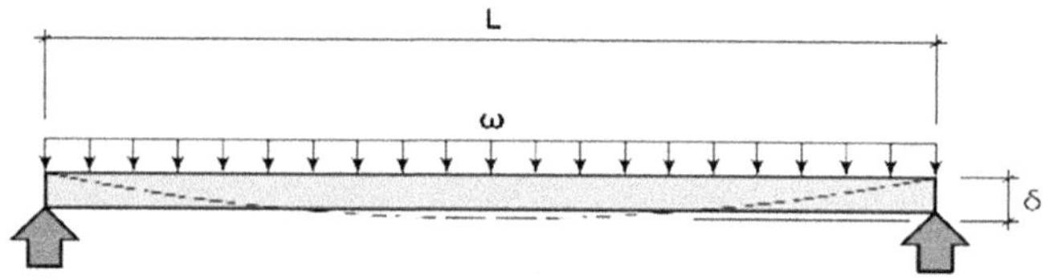

(a) Under a Uniform Load

Maximum deflection under a point load at midspan (Figure 4.7b):

$$\boldsymbol{\delta = PL^3 / 48\ EI}$$

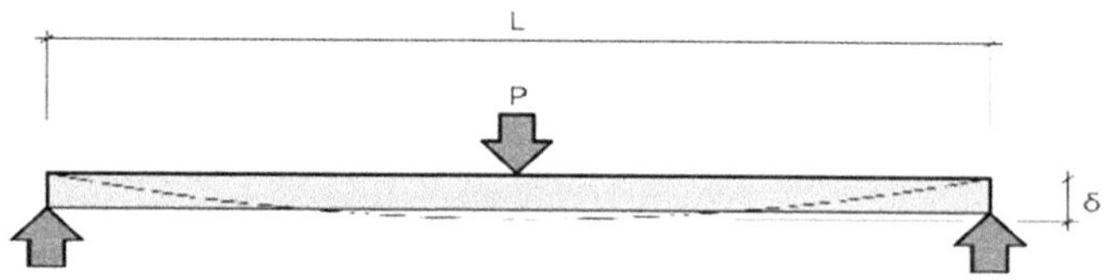

(b) Under a Point Load at Midspan

Maximum deflection under three equally spaced point loads at the quarter points (Figure 4.7c):

$$\boldsymbol{\delta = 0.05\ PL^3 / EI}$$

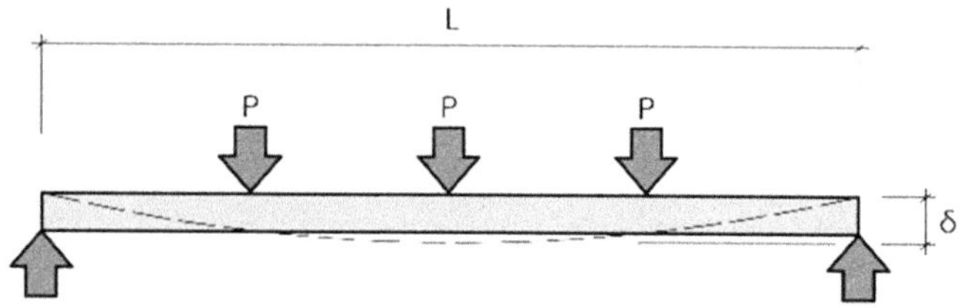

(c) Under 3 Equally Space Point Loads

where:

δ = maximum deflection (in)
w = uniform (service) load (lb/ft or k/ft)
wL = total load (lb or k)
L = length (in)
P = point (service) load (lb or k)
E = modulus of elasticity (lb/in^2 or k/in^2)
I = moment of inertia (in^4)

Figure 4.7 Maximum Deflection Formulae for Simply Supported Beams

Allowable Deflection Formulae

Figure 4.8 shows the code-recommended allowable deflection limits for floor beams.

- For live load (Figure 4.8a):

Deflection may not exceed 1/360 of the span:

$$\Delta_{LL} = L / 360$$

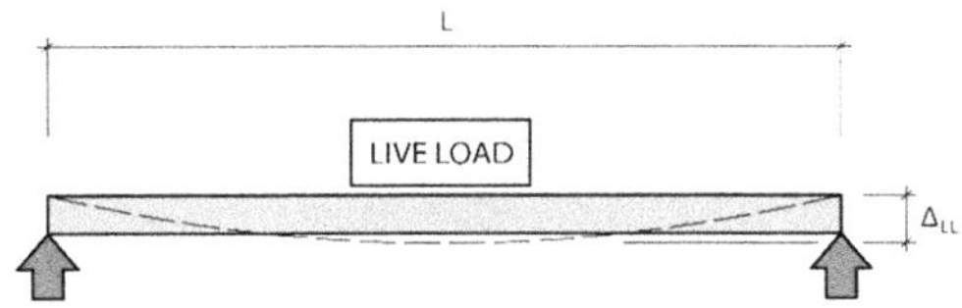

(a) For Live Load Only

- For total (dead + live) load (Figure 4.8b):

Deflection may not exceed 1/240 of the span:

$$\Delta_{TL} = L / 240$$

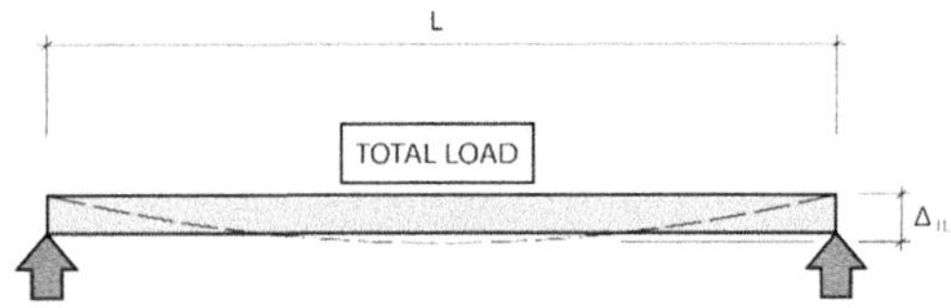

(b) For Total (Live + Dead) Load

where:

Δ_{LL} = live load allowable deflection (in)
Δ_{TL} = total load allowable deflection (in)
L = length (in)

Figure 4.8 Allowable Deflection

Lateral Bracing of Beams

Similar to the tendency of a column to buckle in compression, a beam subject to bending also has the tendency to laterally deform and buckle along the compression edge.

For example, the simply supported beam in Figure 4.9 has its top edge in compression and its bottom edge in tension, thereby tending to buckle laterally along the top edge.

To prevent lateral buckling, the beam must be laterally braced. This can typically be accomplished by additional framing members that brace the compression flange, or by proper fastening to the floor deck materials.

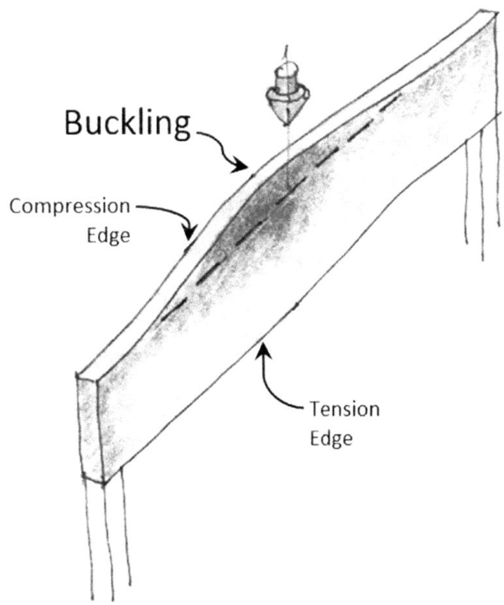

Figure 4.9 Lateral Buckling of a Beam

Beam Penetrations

Beam penetrations are sometimes required to allow for utility and mechanical penetrations (Figure 4.10), and should be minimized to the degree possible.

When penetrations are required, areas of high stress should be avoided such as those close to the supports (where shear force is at a maximum) and those at midspan (where bending stress is at a maximum). For a simply supported beam with symmetrical loads, the desired location is along the neutral axis at the approximate 1/3 points of the span (Figure 4.11).

Figure 4.10 Penetration in Web of a Beam

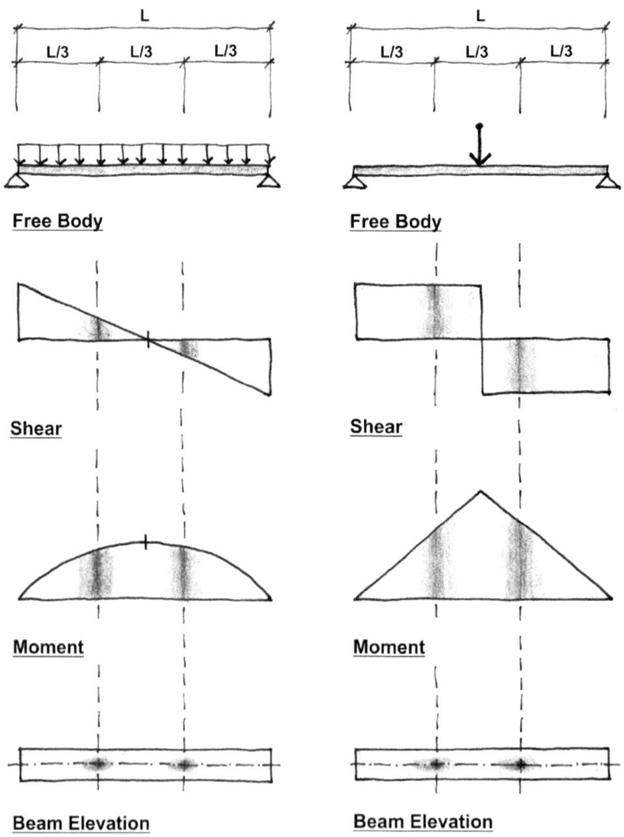

Figure 4.11 Preferred Zones for Penetrations in a Simply Supported Beam

4.3 DESIGNING FOR COLUMNS

A column is typically a vertical member primarily in compression, but may also be subject to bending from lateral and/or eccentric loading. Its design is dependent upon its material, cross-sectional properties, loads, length and end restraints, and any lateral supports.

In earlier times columns tended to be massive and therefore stable under their own weight. With the development of iron, steel, and reinforced concrete, columns become more slender in form. A column's proportions, as defined by its slenderness ratio, is an important consideration which greatly affects its ability to support load.

Effective Length

A column's effective length (L_e) is an important concept in determining its slenderness ratio, and is a function of its actual length (L) and its end restraints (as determined by a k-factor). It is given by:

effective length (L_e) = kL

Slenderness Ratio

Slenderness ratio refers to the proportional relationship between a column's effective length and its least radius of gyration. It is given by:

slenderness ratio = L_e / r where:

L = actual length of column
k = k-factor (a condition of the column's end restraints)
L_e = kL = effective length
r = least radius of gyration (a measure of the column's cross-sectional stiffness)

The Reader is referred to basic structural texts for additional information on k-factor and radius of gyration.

Slenderness ratio affords a means of classifying columns, and is important in design considerations. When overstressed, shorter columns tend to fail by crushing (Figure 4.12), while longer columns tend to fail by buckling (Figure 4.13). The greater the slenderness ratio, the greater the tendency to buckle. The load at which a column fails, either by crushing or buckling, is termed its *critical load*.

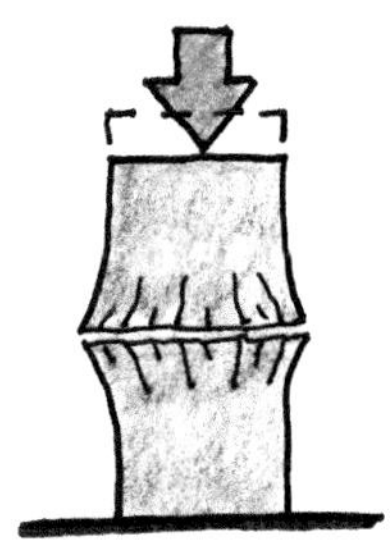

Figure 4.12 Column Crushing

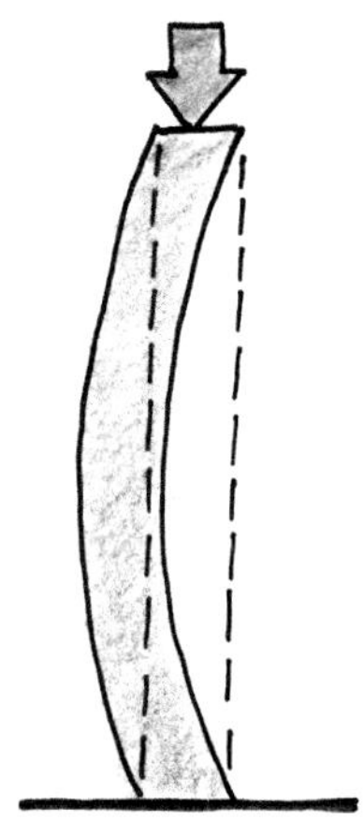

Figure 4.13 Column Buckling

Column Compression and Crushing

The critical load for shorter columns tending to fail by crushing is given by:

$$\mathbf{P_{cr} = f_{cr} \times A}$$

or, in terms of critical stress:

$$\mathbf{f_{cr} = P_{cr} / A}$$

where:

P_{cr} = critical load at which the column will crush
f_{cr} = critical stress at which the column will crush
A = area of cross section

Column Bending and Buckling

As a column's length increases from short to long, there comes a point at which failure by buckling becomes a greater concern than failure by crushing. The critical load for longer columns tending to fail by buckling is given by Euler's formula:

$$\mathbf{P_{cr} = \pi^2 EI / (kL)^2}$$

or, in terms of critical stress:

$$\mathbf{f_{cr} = \pi^2 E / (kL/r)^2}$$

where:

P_{cr} = critical load at which the column will buckle
f_{cr} = critical stress at which the column will buckle
E = modulus of elasticity
I = least moment of inertia
k = k-factor
L = column length
r = least radius of gyration

The Standard Column Curve

The Standard Column Curve represents the relationship between a column's critical stress (f_{cr}) and slenderness ratio (kL/r). The generalized curve in Figure 4.14 shows that as the slenderness ratio increases, the critical stress (and critical load) decreases.

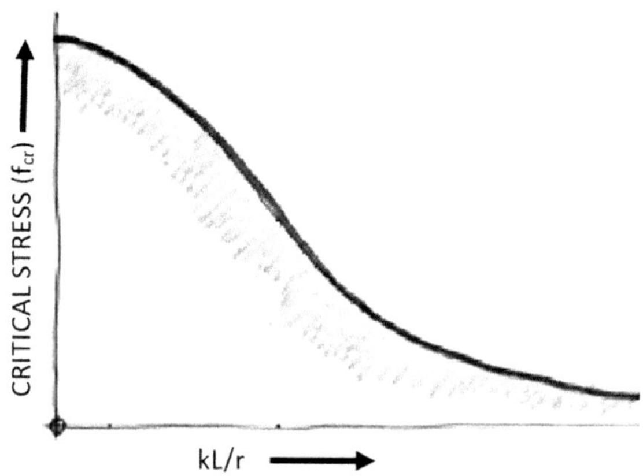

Figure 4.14 The Standard Column Curve

Case Study Assumptions

For simplicity, Case Study columns will be considered to have:

- axial loads only, not subject to lateral loads
- pinned top and bottom restraints; therefore, the k-factor = 1

General Design Approach

The general design approach for a column varies with material as will be shown in the Case Studies, but the concepts of slenderness ratio, critical load, and critical stress are fundamental to column design in all materials. The dilemma often encountered is that a column's slenderness ratio is dependent upon its cross-sectional shape—but we won't know its shape until it's designed. In wood and reinforced concrete, this is resolved by assuming an initial cross section and then verifying the assumption at the conclusion of design. In steel, the AISC simplifies the selection of columns by the use of tables that take the member's cross-sectional properties into account.

We're now ready to proceed with the Case Studies. Prior to each Case Study will be a chapter focused on understanding the subject material's unique characteristics.

5 Understanding Steel

5.1 MANUFACTURE AND MATERIALS

Steel, an alloy composed primarily of iron and carbon, is an extremely strong and durable material. Controlling the carbon content (which ranges between approximately 0.2% and 1.5% for structural steel) as well as alloying with various other metallic elements are major factors in determining steel's qualities. Too much carbon produces a hard, brittle metal, while too little carbon produces a relatively soft, weak metal. *Structural steel* is the term for steel used in a load-bearing capacity (Figure 5.1).

Figure 5.1 A Steel Frame Building

Manufacture

Iron, the predominant ingredient in steel, is a basic element naturally present in certain rock formations. This iron-rich rock is termed iron ore (Figures 5.2 and 5.3).

Figure 5.2 Mining for Iron Ore

Figure 5.3a Stockpiling Iron Ore

Figure 5.3b

Iron is extracted from the ore by the blast furnace process and cast into pigs (large iron bricks with high carbon content) (Figures 5.4 and 5.5).

Figure 5.4 Blast Furnace

Figure 5.5 Stockpiled Iron Pigs

The pig iron is further processed into steel billets with the desired metallurgical and structural properties, ready for final transformation into structural steel shapes or other uses. (Figures 5.6 and 5.7).

Figure 5.6 Processing Steel Billets

Figure 5.7 Stockpiled Steel Billets

Structural steel is produced in a variety of hot rolled and cold formed sections, also referred to as shapes.

Hot Rolled Structural Steel Sections

Hot rolled sections are produced by passing steel billets at very high temperatures through a succession of rollers that progressively squeeze the billets into the desired configuration (Figure 5.8).

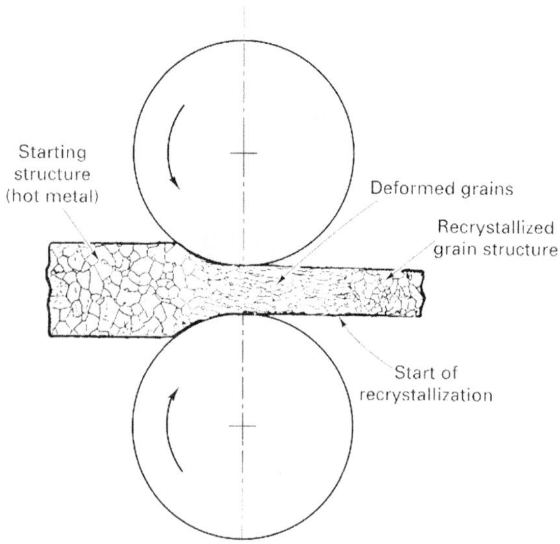

Figure 5.8 Hot Rolling Steel

Common hot rolled shapes are wide flange (W), channel (C), and angle (L) (Figure 5.9).

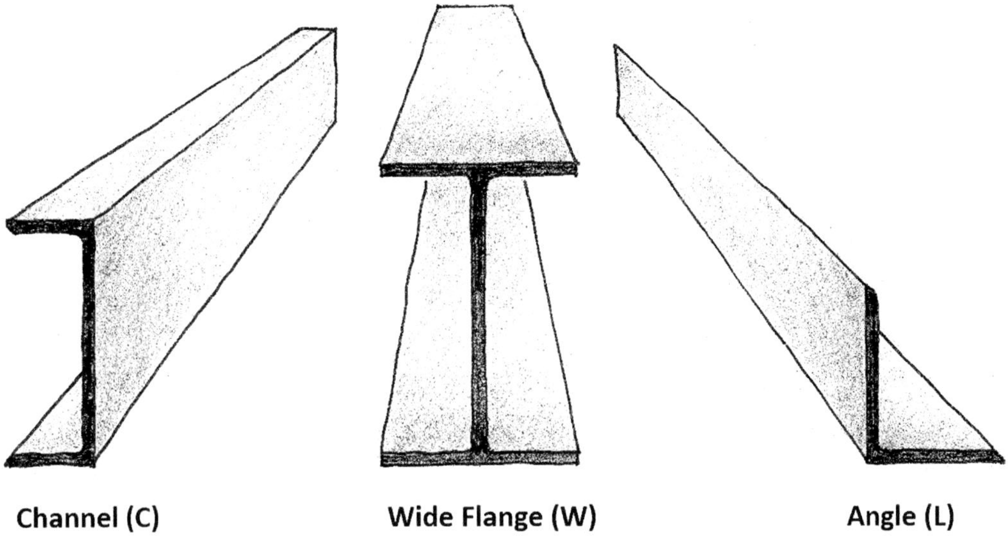

Figure 5.9 Common Hot Rolled Structural Sections

Cold Formed Structural Steel Sections

Cold formed sections are produced by a series of shaping operations at relatively low temperatures that bend sheets of steel of specified thickness into the desired configuration. Common cold formed shapes are hollow structural sections (HSS), circular, square, and rectangular (Figure 5.10).

Figure 5.10 Common Cold Formed Hollow Structural Sections (HSS)

5.2 GENERAL DESIGN CONSIDERATIONS

ASD and LRFD General Strength Equations

As discussed in Chapter 2, the two recognized design methodologies for steel, Allowable Strength Design (ASD) and Load and Resistance Factor Design (LRFD), are separate and distinct approaches, each with their own formulae that must be consistently followed without mixing or matching parts from the other.

The ASD and LRFD general strength equations for the relationship between available strength and required strength are:

ASD

$$\mathbf{R_a \leq R_n / \Omega}$$

where:

R_a = required strength
R_n / Ω = available strength (termed "allowable strength" in ASD)
R_n = nominal strength
Ω = safety factor

LRFD

$$\mathbf{R_u \leq \Phi R_n}$$

where:

R_u = required strength
ΦR_n = available strength (termed "design strength" in LRFD)
R_n = nominal strength
Φ = resistance factor

Values for the ASD safety factor (Ω) and the LRFD resistance factor (Φ), for various conditions, are given in the applicable chapters of the AISC Specification and also appear in the AISC design tables in Appendix 1. Although termed differently, keep in mind that (Ω) and (Φ) both serve as safety factors in the general sense of the word.

The Reader will also note that ASD "allowable strength", and LRFD "design strength", both serve as an "available strength" and will be so referenced for conceptual consistency.

Grades and Strength of Steel

Various shapes are manufactured in different grades of steel, with varying yield strengths (F_y). Although steel of higher yield strengths are available, the most common grades are given by the ASTM designations A36, A992, A53, and A500 (Table 5.1).

Table 5.1 Common Grades of Structural Steel

Grade	Yield Strength (F_y)	Availability
A36	36 ksi	C, L, S, MC shapes
A992	50 ksi	W shapes
A500	46 ksi 50 ksi	Circular HSS shapes Rectangular HSS shapes

Modulus of Elasticity

Steel's high modulus of elasticity makes it very stiff, providing a high resistance to bending in beams and buckling in columns. For commonly used hot rolled and cold formed structural steel, E = 29,000 ksi.

5.3 DESIGN CONSIDERATIONS FOR BEAMS

Resisting Moment and Shear—ASD / LRFD Strength Equations

ASD

Resisting Moment

For moment, the strength equation is written as:

$$\mathbf{M_a \leq M_{px} / \Omega_b}$$

where:

M_a = required flexural strength (i.e., maximum moment)

M_{px} / Ω_b = available flexural strength

M_{px} = nominal flexural strength

Ω_b = safety factor in bending = 1.67

Determination of Z_x

A variation of the flexure formula (F = M / Z) is used to determine the required section modulus.

$$\mathbf{Z_x \geq M_{px} / F_y}$$

or

$$\mathbf{Z_x \geq (M_a \times \Omega_b) / F_y}$$

where:

Z_x = required section modulus

$M_{px} = (M_a \times \Omega_b)$

F_y = yield stress

Resisting Shear

For shear, the strength equation is written as:

$$\mathbf{V_a \leq V_{nx} / \Omega_v}$$

where:

V_a = required shear strength (i.e., maximum shear)

V_{nx} / Ω_v = available shear strength

V_{nx} = nominal shear strength

Ω_v = safety factor in shear = 1.50

LRFD

Resisting Moment

For moment, the strength equation is written as:

$\mathbf{M_u \leq \Phi_b\, M_{px}}$ where:

M_u = required flexural strength (i.e., maximum moment)

$\Phi_b\, M_{px}$ = available flexural strength

M_{px} = nominal flexural strength

Φ_b = resistance factor in bending = 0.90

Determination of Z_x

A variation of the flexure formula (F = M / Z) is used to determine the required section modulus:

$\mathbf{Z_x \geq M_{px} / F_y}$

or

$\mathbf{Z_x \geq M_u / (\Phi_b \times F_y)}$

where:

Z_x = required section modulus

$M_{px} = (M_u / \Phi_b)$

M_u = required flexural strength

F_y = yield stress

Resisting Shear

For shear, the strength equation is written as:

$\mathbf{V_u \leq \Phi_v\, V_{nx}}$ where:

V_u = required shear strength (i.e., maximum shear)

$\Phi_v\, V_{nx}$ = available shear strength

V_{nx} = nominal shear strength

Φ_v = resistance factor in shear = 1.00

Shear in a Wide Flange Beam

Resistance to shear in a wide flange steel beam is essentially provided by the cross-sectional area of the web, or more precisely as shown by Figure 5.11.

$\mathbf{A_w = (d \times t_w)}$ where:

A_w = area of web

d = depth of section

t_w = thickness of web

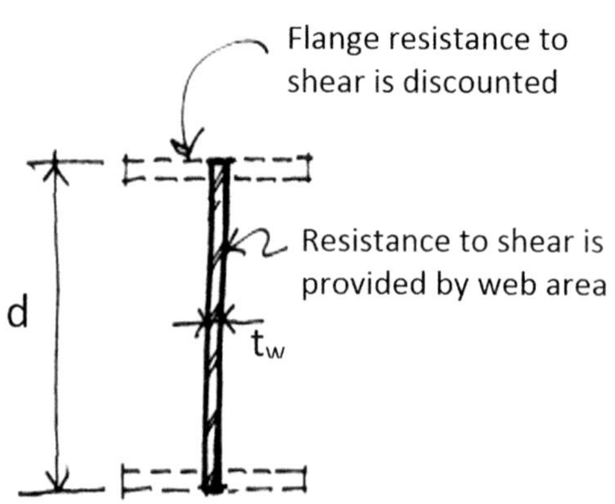

Figure 5.11 Shear Resistance in a W Section

Web Stiffening

In steel construction, when large shear forces are transferred through a beam's web, the shear capacity of the web may need to be increased to prevent the web from buckling. Figure 5.12 shows one method of doing so.

Web buckling is referred to as *web crippling* and the web reinforcing is referred to as *web stiffening*. Web buckling and web stiffening can also be factors in column design.

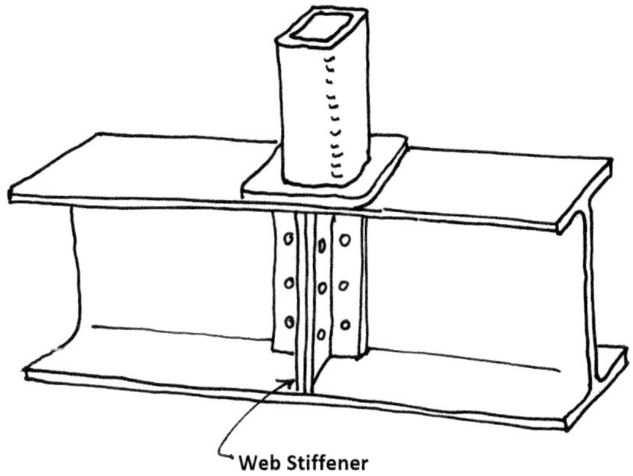

Figure 5.12 Web Stiffener on a Steel Beam

Compact Sections

A compact steel section is one with sufficient code-specified ratios for flange width-to-thickness and web depth-to-thickness, enabling the section to achieve its full flexural strength before experiencing localized buckling failure. The AISC specifies equations for limiting these ratios based on the E and F_y of the material.

Not all steel sections are compact, and the AISC so identifies them. When using non-compact sections, other factors such as web crippling and torsional action may need to be considered. Most commonly used W sections are compact (Figure 5.13).

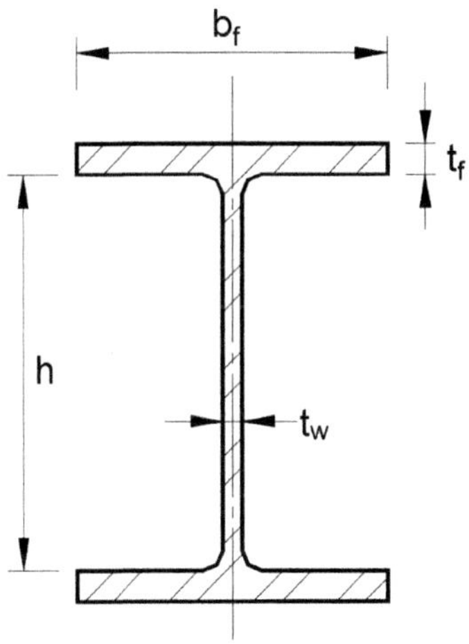

Figure 5.13 Flange and Web Thickness Ratios for a W Compact Section

5.4 DESIGN CONSIDERATIONS FOR COLUMNS

Resisting Compression—ASD / LRFD Strength Equations

For axial compression, the strength equation is written as:

ASD

$$\mathbf{P_a \leq P_n / \Omega_c}$$

where:

P_a = required compressive strength (i.e., loads)
P_n / Ωc = available compressive strength
P_n = nominal compressive strength
Ω_c = safety factor in compression = 1.67

LRFD

$P_u \leq \Phi_c P_n$ where:

P_u = required compressive strength (i.e., loads)
$\Phi_c P_n$ = available compressive strength
P_n = nominal compressive strength
Φ_c = resistance factor in compression = 0.90

The design of a steel column is simplified by AISC tables that provide a column's available strength based on its effective length with respect to its least radius of gyration. Although we'll use these tables in our Case Study, it is important to understand the theory behind them.

Nominal Compressive Strength

The nominal compressive strength of a steel column is given by:

$P_n = F_{cr} \times A_g$ where:

P_n = nominal compressive strength
F_{cr} = critical stress
A_g = gross area of column section

Column Designations and Critical Stress

Figure 5.14 shows the standard column curve in steel. It shows:

- Short steel columns (tending to fail by crushing) are considered to have slenderness ratios less than:
  $4.71 \times \sqrt{(E/F_y)}$
- Long steel columns (tending to fail by buckling) are considered to have slenderness ratios greater than:
  $4.71 \times \sqrt{(E/F_y)}$

Note that the AISC does not recommend using columns with slenderness ratios greater than 200.

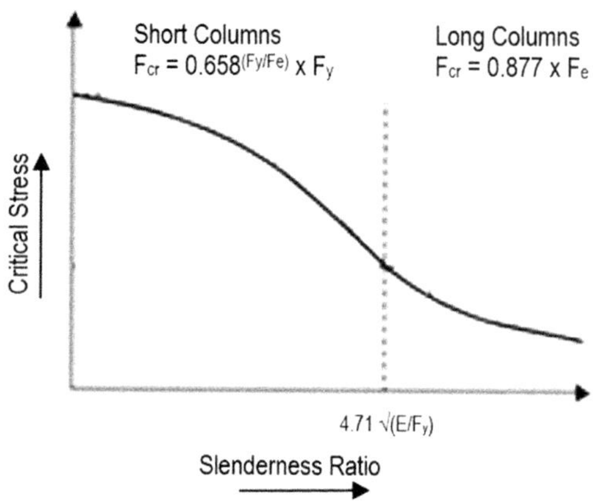

Figure 5.14 Standard Column Curve for Steel

The critical stress of a steel column is given by:

For short columns with	**$kL/r \leq 4.71 \times \sqrt{(E/F_y)}$:**	**$F_{cr} = [0.658^{(Fy/Fe)}] \times F_y$**
For long columns with	**$kL/r > 4.71 \times \sqrt{(E/F_y)}$:**	**$F_{cr} = 0.877 \times F_e$**

In the equations above, F_e is given by Euler's formula: **$F_e = \pi^2 E / (kL/r)^2$.**

AISC Table 4-14 lists the critical stresses for slenderness ratios varying from 1 to 200 for various grades of steel.

Column Shapes

HSS shapes are generally suitable for columns with light to moderate loads. W shapes are generally suitable for columns with heavier loads such as multistory buildings. Since W shapes allow for simpler connections to beams and girders, they may often be used for columns with lighter loads as well.

Column Orientation

At first glance it may seem odd that a W shape would be used both for a beam in bending and a column in compression. However, when column buckling (i.e., bending) is considered, the rationale becomes apparent.

For structural efficiency, a column is best oriented to provide its greatest resistance to bending (i.e., maximum stiffness) in the direction of lateral force. For example, the W column in Figure 5.15 has a greater moment of inertia (I) about its xx axis than its yy axis. For the given lateral force, the column is therefore best oriented as shown to provide its maximum resistance to bending about the xx axis.

The steel framing plan in Figure 6.1 (Chapter 6 Case Study) shows that the loading on interior columns is symmetrical—therefore their orientation is not critical. However, since the loading on perimeter columns is asymmetrical, they are oriented to provide the most efficient resistance to bending.

Column orientation is especially important when a column is a part of a lateral resistance system.

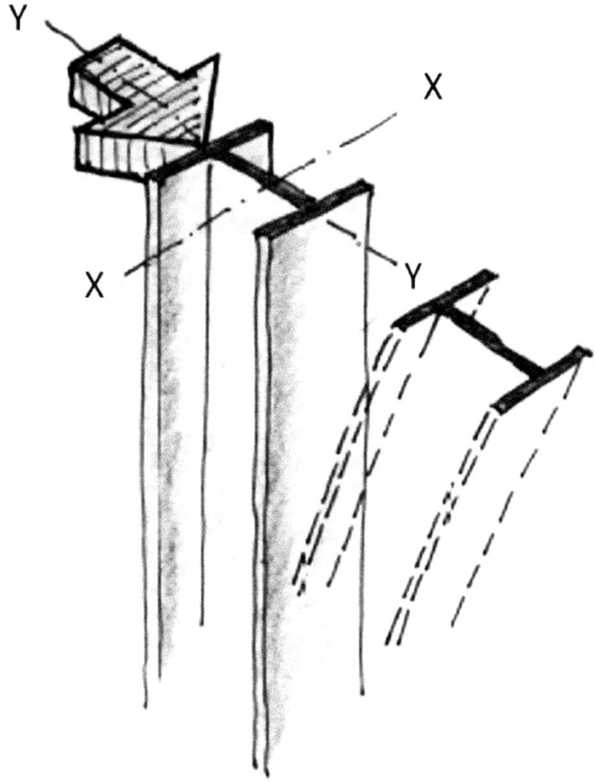

Figure 5.15 Orientation of a W Column for Maximum Resistance to Bending

6 Design in Steel—Case Study

Our Case Study for design in steel is a two-story one-way framed structure, 48 ft × 48 ft, with beams spanning 24 ft, girders spanning 16 ft, and 12 ft floor-to-floor heights. We'll focus on the structural design of the first floor and select typical members Beam 3, Girder B, and Column B2 to design. In addition to the loads from the first floor, the loads from the roof will be added onto Column B2. Figure 6.1 shows the first floor framing plan.

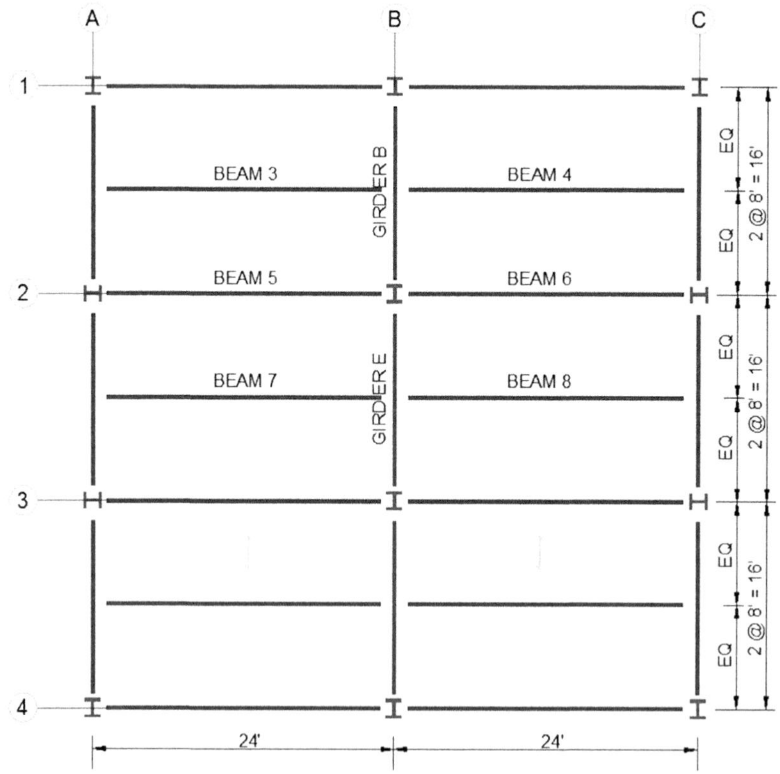

Figure 6.1 First Floor Framing Plan

We'll design the typical members using Allowable Strength Design (ASD) and Load and Resistance Factor Design (LRFD). For either methodology, the AISC Manual contains many approaches, tables, and charts from which we can select structural members and perform checks (see Appendix 1). For our Case Study, we've chosen one commonly used approach for selecting beams based on their section modulus (*Zx*).

6.1 ASSUMPTIONS

Floor Construction

Floor construction will be nominal 3″ of lightweight concrete, on a 3″ 20-gage metal deck, spanning the steel beams as a one-way slab (Figure 6.2).

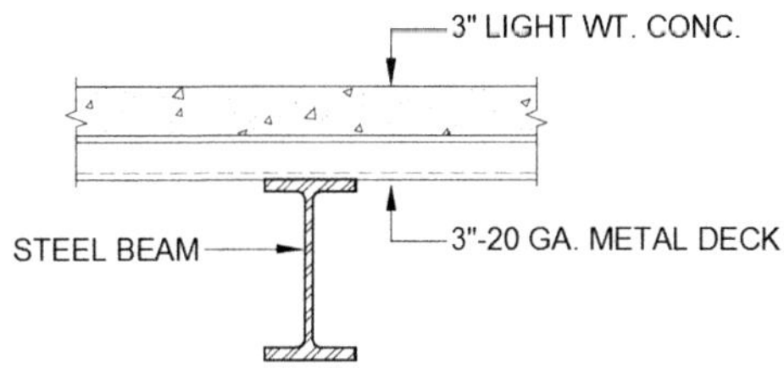

Figure 6.2 Floor Construction

Member Shapes and Material Properties

Beams and girders will be W shapes since they are generally the most efficient for these types of members. For columns, we'll evaluate both W and HSS shapes. Relevant material properties for the Case Study are:

For W Shapes	For Rectangular HSS Shapes
▪ grade of steel: A992 ▪ yield stress (F_y): 50 ksi	▪ grade of steel: A500 Grade C ▪ yield stress (F_y): 50 ksi
▪ modulus of elasticity (E): 29,000 ksi	

Floor and Roof Loads

We'll assume the following service loads:

Floor Loads		Roof Loads	
Dead Load		***Dead Load***	
3″ lightweight concrete	= 45 psf	roofing and insulation	= 10 psf
3″ 20-gage metal deck	= 2 psf	roof metal deck	= 2 psf
*self-weight of steel members	= 10 psf	*self-weight of steel members	= 8 psf
ceiling, floor finishes	= 10 psf	ceiling finish	= 2 psf
mechanical (pipes, ducts, etc.)	= 8 psf	mechanical (pipes, ducts, etc.)	= 8 psf
Total	**= 75 psf**	**Total**	**= 30 psf**
Live Load		***Live Load***	
Total	**= 75 psf**	**Total**	**= 30 psf**

*In steel design, the self-weight of structural members is relatively small in relation to the other dead loads, and is typically included as a suitable square foot allowance in the dead load calculations.

AISC Tables

For various AISC tables in Appendix 1, ASD tabular columns are shown shaded, while LRFD tabular columns are shown unshaded.

Beams and Girders

Beams and girders will be considered to have:

- simple supports
- full lateral bracing

Figure 6.3 shows the typical steps we'll follow in the design of Beam 3 and Girder B.

1. design to resist moment
2. check for shear
3. check for deflection

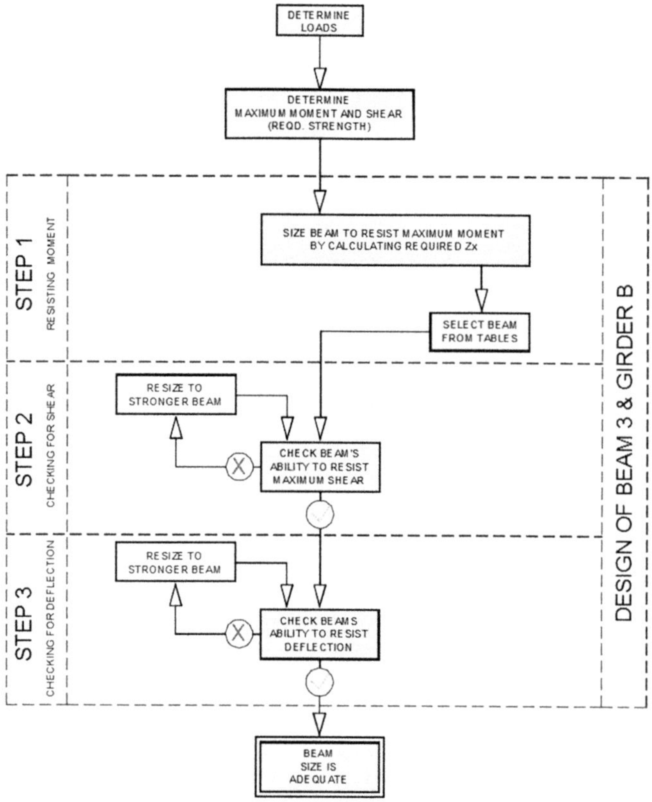

Figure 6.3 Flow Chart for Design of Beam 3 and Girder B

Columns

Columns will be considered to have:

- axial loads only, not subject to lateral loads
- floor-to-floor height (L) = 12 ft
- pinned top and bottom restraints; therefore, k-factor = 1
- effective length $L_e = kL = 12$ ft

Figure 6.4 shows the typical steps we'll follow in the design of Beam 3 and Girder B.

Let's proceed to design our typical members, first using Allowable Strength Design (ASD) and then using Load and Resistance Factor Design (LRFD).

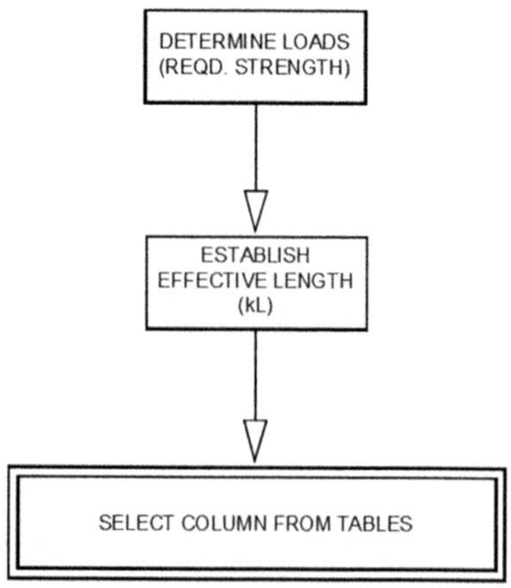

Figure 6.4 Flow Chart for Design of Column B2

CASE STUDY—DESIGN IN STEEL (ASD)

6.2.ASD BEAM 3

Determine Loads (Beam 3—ASD)

Calculate the uniform load imposed on Beam 3.

The load tributary area for Beam 3 is a section of floor 8 ft × 24 ft (Figure 6.5A). The uniformly distributed loads (w) on Beam 3 are:

$$\begin{aligned}
w_{DL} &= \text{uniform dead load} \\
&= 75\ \text{lb/ft}^2 \times 8\ \text{ft} = 600\ \text{lb/ft} = 0.60\ \text{k/ft} \\
w_{LL} &= \text{uniform live load} \\
&= 75\ \text{lb/ft}^2 \times 8\ \text{ft} = 600\ \text{lb/ft} = 0.60\ \text{k/ft} \\
\hline
w_{TL} &= \text{uniform total (service) load} = 1.20\ \text{k/ft}
\end{aligned}$$

Applying the governing ASD load combination (1.0 D + 1.0 L) to $w_{DL} + w_{LL}$ (see Chapter 4):

Total ASD Design Load (w_a)

$$\begin{aligned}
\mathbf{w_a} &= 1.0 w_{DL} + 1.0 w_{LL} \\
&= (1.0 \times 0.60) + (1.0 \times 0.60) \\
&= \mathbf{1.20\ k/ft}
\end{aligned}$$

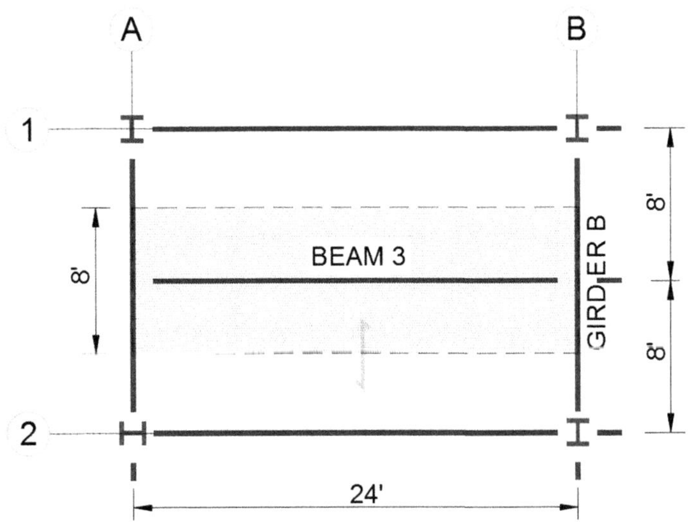

Figure 6.5A Load Tributary Area of Beam 3

Determine Reactions, Maximum Shear, Maximum Moment (Beam 3—ASD)

Beam 3 is simply supported with a uniformly distributed load along its length. From Appendix 4—Beam Diagrams and Formulae—the shear and moment diagrams with maximum values for Beam 3 are shown in Figure 6.6A.

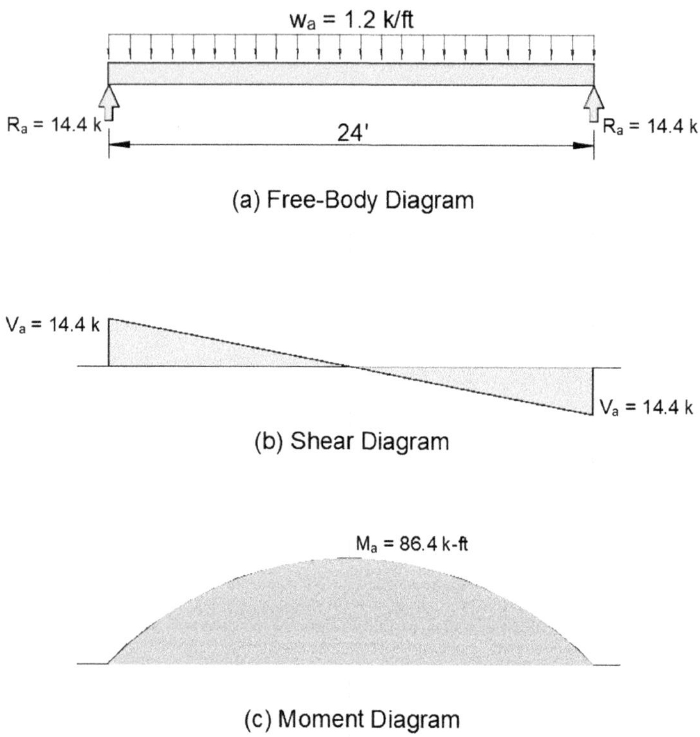

Figure 6.6A Beam 3: Free-body, Shear & Moment Diagrams

$L\ (\text{length}) = 24\ \text{ft}$

$w_a = 1.2\ \text{k/ft}$

Reactions (R_a)

$R_a = w_a L/2$

$R_a = (1.2 \times 24)/2$

$\mathbf{R_a = 14.4\ k}$

Maximum Shear (V_a)

$V_a = R_a$

$\mathbf{V_a = 14.4\ k}$

(Required Shear Strength)

Maximum Moment (M_a)

$M_a = wL^2/8$

$M_a = (1.2 \times 24^2)/8$

$\mathbf{M_a = 86.4\ k\text{-}ft}$

(Required Flexural Strength)

Design of Beam 3 (ASD)

Step 1: Resisting Moment (Beam 3—ASD)

Designing to resist moment involves selecting a beam with an available flexural strength equal to or greater than the required flexural strength—as expressed by:

$$\mathbf{M_a \le M_{px} / \Omega_b}$$

where:

M_a = required flexural strength

M_{px} / Ω_b = available flexural strength

Calculate the Required Section Modulus (Z_x)

$$\mathbf{Z_x = (M_a \times W_b) / F_y}$$
$$= [(86.4 \times 12) \times (1.67)] / 50$$
$$Z_x = 34.6 \text{ in}^3$$

where:

Z_x = required section modulus (in^3)
M_a = maximum moment = 86.4 k-ft
Ω_b = safety factor = 1.67
F_y = yield stress = 50 k/in^2

(for unit consistency, M_a is multiplied by 12 to convert feet to inches.)

Select the Beam from Tables

From AISC Table 3-2 (W-Shapes—Selection by Z_x), select a shape with an available Z_x equal to or greater than that required. Note that we are looking for the lightest and therefore most economical member that satisfies the design requirement.

A W8x35 shape, with an available Z_x of 34.7 in^3, is the closest (but still greater) value to the required Z_x of 34.6 in^3. Note that this shape's available flexural strength of 86.6 k-ft is greater than the required flexural strength of 86.4 k-ft.

Shape	Available Section Modulus (Z_x)	Required Section Modulus (Z_x)
W8×35	34.7 in^3	34.6 in^3

Available Flexural Strength (M_{px} / Ω_b)	Required Flexural Strength (M_a)
86.6 k-ft	86.4 k-ft

While the W8×35 has an available Z_x greater than that required, note that it is not the lightest shape to satisfy the design requirement. The deeper W12×26 at the top of the group in the AISC table, having a lighter weight (26 plf) and higher available Z_x (37.2 in^3), is given in boldface indicating that it is the lightest shape of the group that will satisfy the design requirement. Also, note that the available flexural strength of the W12×26 is 92.8 k-ft, greater than the required flexural strength of 86.4 k-ft.

Shape	Available Section Modulus (Z_x)	Required Section Modulus (Z_x)
W12×26	37.2 in^3	34.6 in^3

Available Flexural Strength (M_{px} / Ω_b)	Required Flexural Strength (M_a)
92.8 k-ft	86.4 k-ft

We'll tentatively select the W12×26 (with an I_x of 204 in^4) for Beam 3 pending a shear and deflection check.

Step 2: Checking for Shear (Beam 3—ASD)

Checking for shear involves assuring that a beam's available shear strength is equal to or greater than its required shear strength, as expressed by:

$$\mathbf{V_a \le V_{nx} / \Omega_v}$$

where:
V_a = required shear strength
V_{nx} / Ω_v = available shear strength

From AISC Table 3-2, the available shear strength of a W12×26 is 56.1 k.

From the shear diagram, the required shear strength is 14.4 k.

Shape	Available Shear Strength (V_{nx} / Ω_v)	Required Shear Strength (V_a)
W12×26	56.1 k	14.4 k

Since the available shear strength is considerably greater than the required shear strength, the W12×26 passes the check for shear.

AS A MATTER OF INTEREST:
Calculating ASD Available Shear Strength in a W Beam

Having checked the available shear strength of the W12x26 beam using AISC tables, let's verify it using the formulae in Chapter 5 (refer to Figure 5.11).

$$\mathbf{V_a \le V_{nx} / \Omega_v}$$

The AISC provides the following formula for V_{nx}:

$$\mathbf{V_{nx} = 0.6\ F_y \times A_w \times C_w}$$

Therefore:

$$V_a = (0.6\ F_y \times A_w \times C_w) / \Omega_v$$
$$= (0.6 \times 50 \times 2.81 \times 1) / 1.50$$
$$= 56.2\ \text{k}$$

where:
V_a = required shear strength
V_{nx} / Ω_v = available shear strength
V_{nx} = nominal shear strength
Ω_v = safety factor = 1.50
F_y = yield stress = 50 k/in^2
A_w = area of web = d × t_w = 12.2 × 0.23 = 2.81 in^2 (from AISC Table 1-1):
d = depth of web = 12.2 in
t_w = thickness of web = 0.23 in
C_w = web shear coefficient for beams (given as 1.0 for W beams)

We see that the 56.2 k available shear strength of the W12x26 obtained from formula calculations is the same as that obtained from AISC tables.

Step 3: Checking for Deflection (Beam 3—ASD)

For the 24 ft long W12×26 beam, assure that its maximum deflection is less than or equal to its allowable deflection for both:

- live load (only)
- total load (dead + live)

Live Load Deflection

- Live Load: Maximum Deflection for the W12×26 beam

$$\boldsymbol{\delta_{LL} = 5(w_{LL}L \times L^3) / 384\ EI}$$

where:

δ_{LL} = live load maximum deflection (in)
w_{LL} = uniform live load = 0.60 k/ft
L = length = 24 ft
E = modulus of elasticity = 29,000 k/in^2
I_x = moment of inertia = 204 in^4

$$\delta_{LL} = \left[(5) \times (0.6 \text{ x } 24) \times (24 \times 12)^3\right] / (384 \times 29{,}000 \times 204) = 0.76 \text{ in}$$

- Live Load: Allowable Deflection for the W12×26 beam

$$\boldsymbol{\Delta_{LL} = L / 360}$$

where:

Δ_{LL} = live load allowable deflection (in)

$$\Delta_{LL} = (24 \times 12) / 360 = 0.80$$

Shape	Live Load Maximum Deflection (δ_{LL})	Live Load Allowable Deflection (Δ_{LL})
W12×26	0.76 in	0.80 in

Since the live load maximum deflection is less than the live load allowable deflection, the W12×26 passes the check for live load deflection.

Total Load Deflection

- Total Load: Maximum Deflection for the W12×26 beam

$$\boldsymbol{\delta_{TL} = 5(w_{TL}L \times L^3) / 384\ EI}$$

where:

δ_{TL} = total load maximum deflection (in)
w_{TL} = uniform total load = 1.20 k/ft
L = length = 24 ft
E = modulus of elasticity = 29,000 k/in^2
I_x = moment of inertia = 204 in^4

$$\delta_{TL} = \left[(5) \times (1.2 \times 24) \times (24 \times 12)^3\right] / (384 \times 29{,}000 \times 204) = 1.51 \text{ in}$$

- Total Load: Allowable Deflection for the W12×26 beam

$$\Delta_{TL} = L / 240$$

where:

Δ_{TL} = total load allowable deflection (in)

$$\Delta_{TL} = (24 \times 12) \ / \ 240 = 1.20 \text{ in}$$

Shape	Total Load Maximum Deflection (δ_{TL})	Total Load Allowable Deflection (Δ_{TL})
W12×26	1.51 in	1.20 in

Since the total load maximum deflection is greater than the total load allowable deflection, the W12×26 does NOT pass the check for total load deflection.

Return to Table 3-2 and select the next higher shape (i.e., the shape in bold font having the next greater Z_x), W14×26 with an I_x of 245 in^4. The total load maximum deflection for this beam is 1.26 in—still greater than the total load allowable deflection (1.20 in).

Return again to Table 3-2 and select the next larger economical shape (in bold font) W16×26 with an I_x of 301 in^4.

- Total Load: Maximum Deflection for the W16×26 Beam

$$\delta_{TL} = \left[(5) \times (1.2 \times 24) \times (24 \times 12)^3\right] / (384 \times 29{,}000 \times 301) = 1.02 \text{ in}$$

Shape	Total Load Maximum Deflection (δ_{TL})	Total Load Allowable Deflection (Δ_{TL})
W16x26	1.02 in	1.20 in

Since the total load maximum deflection is less than the total load allowable deflection, the W16×26 passes the check for total load deflection, so we'll confirm it as our selection for Beam 3.

AS A MATTER OF INTEREST: Beam 3 Selection from Tables

The AISC provides tables for directly selecting a simply supported beam supporting a uniformly distributed load. These tables are based on available flexural and shear strengths.

To select a W16 shape for Beam 3, use AISC Table 3-6 (Maximum Total Uniform Load, kips—W16 Shapes) and convert the uniformly distributed load to a total load.

w = uniformly distributed load = 1.2 k / ft
L = length of beam (i.e., span) = 24 ft

Total Load (i.e., the Maximum Total Uniform Load) = w x L = 1.2 x 24 = 28.8 k.

For a 24-ft span, we see that a W16x26 with a total ASD capacity of 36.8 k is the lightest available W16 shape.

6.3.ASD GIRDER B

For the design of Girder B, we'll follow steps similar to those in Beam 3.

Determine Loads (Girder B—ASD)

Girder B is supporting Beam 3 and Beam 4 at its midspan. The combined design point load (P) from these two identical beams is 28.8 k (i.e., the end reaction of each beam; 14.4 k + 14.4 k) (Figure 6.7A).

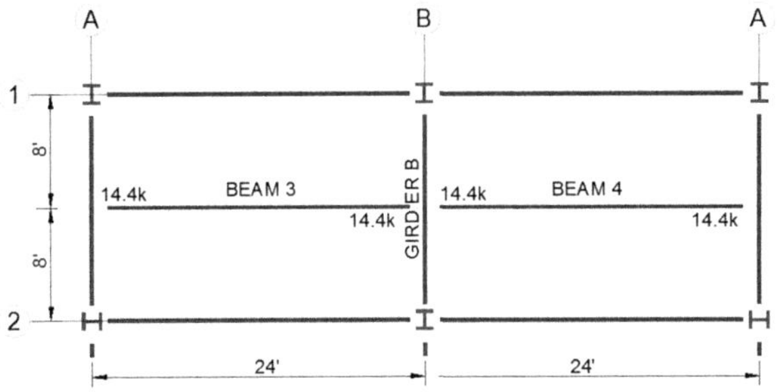

Figure 6.7A Loads on Girder B

Determine Reactions, Maximum Shear, Maximum Moment (Girder B—ASD)

Girder B is simply supported with a point load at its midspan. From Appendix 4—Beam Diagrams and Formulae—the shear and moment diagrams with maximum values for Girder B are shown in Figure 6.8A.

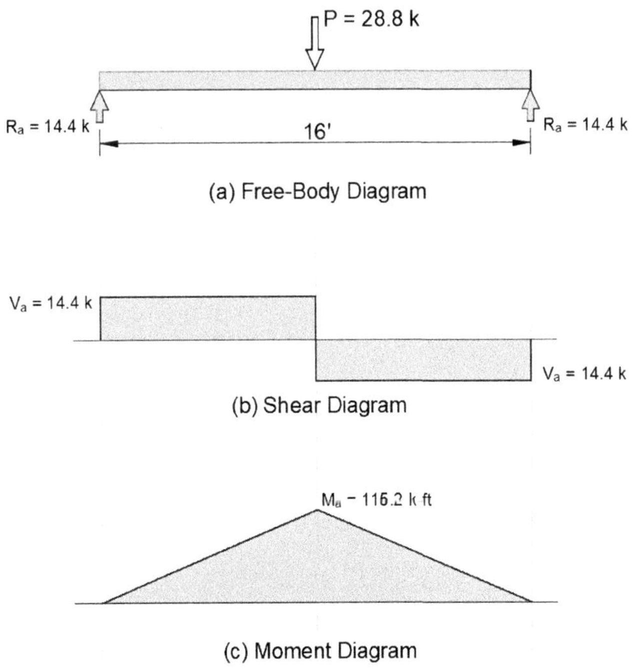

Figure 6.8A Girder B: Free-body, Shear & Moment Diagrams

$L\ (\text{length}) = 16$ ft

$P = 28.8$ k

Reactions (R_a)

$R_a = P/2$

$R_a = 28.8 / 2$

$\mathbf{R_a = 14.4\ k}$

Maximum Shear (V_a)

$V_a = R_a$

$\mathbf{V_a = 14.4\ k}$

(Required Shear Strength)

Maximum Moment (M_a)

$M_a = PL / 4$

$M_a = (28.8 \times 16) / 4$

$\mathbf{M_a = 115.2\ k\text{-}ft}$

(Required Flexural Strength)

Design of Girder B (ASD)

Step 1: Resisting Moment (Girder B—ASD)

Designing to resist moment involves selecting a beam with an available flexural strength equal to or greater than the required flexural strength—as expressed by:

$$\mathbf{M_a \leq M_{px} / \Omega_b}$$

where:
M_a = required flexural strength
M_{px} / Ω_b = available flexural strength

Calculate the Required Section Modulus (Z_x)

$$\mathbf{Z_x = (M_a \times \Omega_b) / F_y}$$
$$= \left[(115.2 \times 12) \times (1.67)\right] / 50$$
(changing k-ft to k-in)
$$Z_x = 46.2 \text{ in}^3$$

where:
Z_x = required section modulus (in^3)
M_a = maximum moment = 115.2 k-ft
Ω_b = safety factor = 1.67
F_y = yield stress = 50 k/in^2

Select the Girder from Tables

From AISC Table 3-2, select a shape with an available Z_x greater than that required.

A W10x39 shape, with an available Z_x of 46.8 in^3, is the closest (but still greater) value to the required Z_x of 46.2 in^3; but we also see that a W14x30 shape, with an available Z_x of 47.3 in^3 is the lightest and therefore most economical shape in this group. Note that the W14×30 available flexural strength of 118 k-ft is greater than the required flexural strength of 115.2 k-ft.

Shape	Available Section Modulus (Z_x)	Required Section Modulus (Z_x)
W14×30	47.3 in^3	46.2 in^3

Available Flexural Strength (M_{px} / Ω_b)	Required Flexural Strength (M_a)
118 k-ft	115.2 k-ft

We'll tentatively select the W14×30, with an I_x of 291 in^4, for Girder B pending a shear and deflection check.

Step 2: Checking for Shear (Girder B—ASD)

For the W14×30 girder, assure that its available shear strength is equal to or greater than its required shear strength, as expressed by:

$$\mathbf{V_a \leq (V_{nx} / \Omega_v)}$$

where:
V_a = required shear strength
V_{nx} / Ω_v = available shear strength

From AISC Table 3-2, the available shear strength of a W14×30 is 74.5 k.

From the shear diagram, the required shear strength is 14.4 k.

Shape	Available Shear Strength (V_{nx} / Ω_v)	Required Shear Strength (V_a)
W14×30	74.5 k	14.4 k

Since the available shear strength is considerably greater than the required shear strength, the W14×30 passes the check for shear.

Step 3: Checking for Deflection (Girder B—ASD)

For the 16 ft long W14×30 girder, assure that its maximum deflection is less than or equal to its allowable deflection for both:

- live load (alone)
- total load (dead + live)

Live Load Deflection

- Live Load: Maximum Deflection for the W14×30 girder

$$\boldsymbol{\delta_{LL} = P_{LL}\, L^3/48\, EI}$$

where:

δ_{LL} = live load maximum deflection (in)

P_{LL} = point live load = 14.4 k

(The total point load is 28.8 k. Since the live and dead loads are equal, the live point load (P) is ½, or 14.4 k.)

L = length = 16 ft

E = modulus of elasticity = 29,000 k/in²

I_x = moment of inertia = 291 in⁴

$$= \left[(14.4) \times (16 \times 12)^3\right] / (48 \times 29{,}000 \times 291) = 0.25 \text{ in}$$

- Live Load: Allowable Deflection for the W14×30 girder

$$\boldsymbol{\Delta_{LL} = L / 360}$$

$$= (16 \times 12) / 360 = 0.53 \text{ in}$$

where:

Δ_{LL} = live load allowable deflection (in)

Shape	Live Load Maximum Deflection (δ_{LL})	Live Load Allowable Deflection (Δ_{LL})
W14×30	0.25 in	0.53 in

Since the live load maximum deflection is less than the live load allowable deflection, the W14×30 passes the check for live load deflection.

Total Load Deflection

- Total Load: Maximum Deflection for the W14×30 girder

$$\delta_{TL} = P_{TL}\, L^3/48\, EI$$

where:

δ_{TL} = total load maximum deflection (in)

P_{TL} = point total load = 28.8 k

$$= \left[(28.8)\times(16\times12)^3\right]/(48\times29{,}000\times291) = 0.50 \text{ in}$$

- Total Load: Allowable Deflection for the W14×30 girder

$$\Delta_{TL} = L\ /\ 240$$
$$= (16\times12)/240 = 0.80 \text{ in}$$

where:

Δ_{TL} = total load allowable deflection (in)

Shape	Total Load Maximum Deflection (δ_{TL})	Total Load Allowable Deflection (Δ_{TL})
W14×30	0.50 in	0.80 in

Since the total load maximum deflection is less than the total load allowable deflection, the W14×30 passes the check for total load deflection, so we'll confirm it as our selection for Girder B.

6.4 ASD COLUMN B2

Determine Loads (Column B2—ASD)

First Floor Loads

Column B2 is supporting first floor Girders B and E and Beams 5 and 6. The combined load from the end reaction of each of these members is tabulated below and shown in Figure 6.9A.

Member	First Floor Load on Column B2
Girder B	14.40 k
Girder E	14.40 k
Beam 5	14.40 k
Beam 6	14.40 k
Total	57.60 k

Roof Loads

In addition to the first floor loads, Column B2 is also supporting roof loads. Since the roof framing and loading is symmetrical about Column B2, we can calculate its roof load based on its load tributary area (Figure 6.10A).

Load tributary area for Column B2 = 16 ft × 24 ft
= 384 sf

D = uniform roof dead load = 30 psf
L_r = uniform roof live load = 30 psf

Applying the governing ASD roof load combination:

$$D + L_r = 30 + 30 = 60\,\text{psf}$$

$$\text{Roof load on Column B2} = 384\text{ft}^2 \times 60\text{lb/ft}^2 = 23{,}040\text{ lb} = 23.04\text{ k}$$

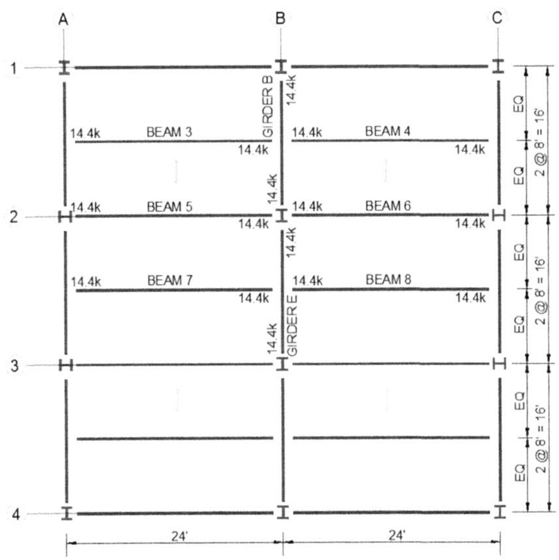

Figure 6.9A First Floor Loads on Column B2

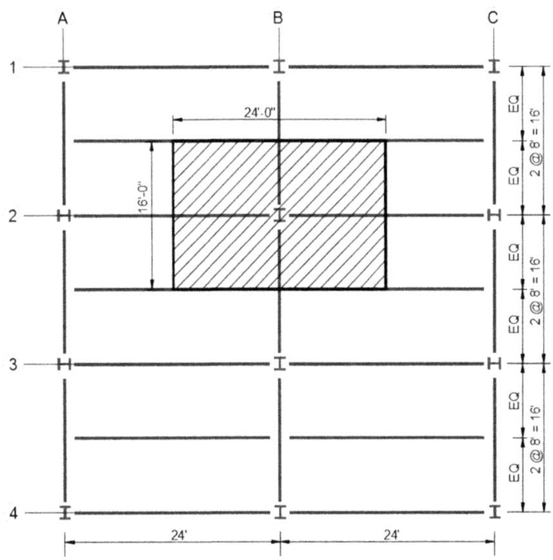

Figure 6.10A Roof Tributary Area for Column B2

Total Floor and Roof Loads on Column B2

First floor load	= 57.60 k
Roof load	= 23.04 k

Total Load on Column B2 = 80.64 k (required compressive strength, P_a)

Establish Effective Length (L_e) (Column B2—ASD)

$L_e = kL = 12$ ft

Design of Column B2 (ASD)

Designing to resist compression involves selecting a column with an available compressive strength equal to or greater than the required compressive strength—as expressed by the strength equation:

$$\mathbf{P_a \leq P_n / \Omega_c}$$

where:

P_a = required flexural strength (i.e., loads)

P_n / Ω_c = available flexural strength

Selecting a W Shape from Tables (Column B2—ASD)

To select a W shape, we'll use AISC Table 4-1a (Available Strength in Axial Compression, kips—for W8 Shapes). This table provides the available strength of various W shapes for a given effective length (L_e) with respect to the shape's least radius of gyration (r), taking into account the column's slenderness ratio and column classification.

For an effective length (L_e) of 12 ft, enter Table 4-1a and select a W8×31 with an available ASD compressive strength (P_n / Ω_c) of 188 k. This is the lightest W shape available for a 12 ft effective length.

Shape	Weight	Available Compressive Strength (P_n / Ω_c)	Required Compressive Strength (P_a)
W8×31	31 lb/ft	189.0 k	80.64 k

We see that the 189.0 k available compressive strength is significantly greater than the 80.64 k required compressive strength.

Selecting a Square HSS Shape from Tables (Column B2—ASD)

As an alternative, let's evaluate a square HSS shape since these are generally used for columns having lighter loads. From AISC Table 4-4 (Available Strength in Axial Compression, kips—for Square HSS Shapes), we'll evaluate several sizes of square HSS sections with an effective length (L_e) of 12 ft:

Shape	Weight (lb/ft)	Available Compressive Strength (P_n / Ω_c)	Required Compressive Strength (P_a)
HSS 4½ × 4½ × 5/16	16.9	82.5 k	80.64 k
HSS 5 × 5 × 1/4	15.6	85.7 k	80.64 k
HSS 5½ × 5½ × 1/4	17.3	102 k	80.64 k
HSS 6 × 6 × 3/16	14.5	91 k	80.64 k

We see that the HSS 6 × 6 × 3/16 is the lightest section with an available compressive strength greater than the required compressive strength.

Discussion

Comparing our two column selections, we see that the weight of the HSS shape is lower than the W shape. As always however, the final selection of an appropriate shape is determined by the designer's judgment based on several criteria such as compatibility with the lateral force resisting system used, the importance and cost of connections, and designer preferences.

AS A MATTER OF INTEREST: Calculating ASD Available Compressive Strength

For the W8x31 ASD column selection, let's verify its available compressive strength (P_a) of 189.0 k using the AISC equations for critical stress.

For the W8x31 (AISC Table 1-1— Dimensions and Properties):

$$k = 1.0 \qquad r_y = 2.02\,in \qquad F_y = 50\,k/in^2$$
$$L = 12\,ft \qquad A_g = 9.13\,in^2 \qquad E = 29{,}000\,k/in^2$$
$$r_x = 3.47\,in$$

Effective Length

$$L_e = kL = 1.0 \times 12 = 12\,ft \times 12 = 144\,in$$

Slenderness Ratio (about yy axis having the least radius of gyration)

$$kL/r_y = (144)/2.02 = 71.29$$

Column Designation

Since: $4.71 \times \sqrt{(E/F_y)} = 4.71 \times \sqrt{(29{,}000/50)} = 113.4$

Column B2 (with a kL/r_y of 71.29) is designated as a short column

Critical Stress (F_{cr})

The critical stress for short columns is given by:

$$F_{cr} = \left[0.658^{(Fy/Fe)}\right] \times F_y$$
$$= \left[0.658^{(50/56.32)}\right] \times 50 = 34.48 \text{ k / in}^2$$

where:

$F_e = \pi^2 E / (kL/r)^2$
$= (3.14^2 \times 29{,}000) / (71.29)^2$
$= 56.32 \text{ k/in}^2$

Nominal Compressive Strength (P_n)

$$P_n = F_{cr} \times A_g$$
$$= 34.48 \times 9.13$$
$$= 314.82 \text{ k}$$

Available Compressive Strength (P_n / Ω_c)

$P_n / \Omega_c = 314.82 / 1.67$ where: $\Omega_c = 1.67$
$= 188.53 \text{ k}$

We see that the available compressive strength of the W8x31 column by calculations is virtually the same as the 189 k obtained from AISC Table 4-1a.

Note: AISC Table 4-14 (Available Critical Stress for Compression Members) provides the available critical stress for various 'L_e/r' ratios (F_{cr} / Ω_c and Φ_c F_{cr} in ASD and LRFD respectively). For a 'L_e/r' of 71.29, we see from the table (by interpolation) that the ASD available critical stress is 20.6 ksi. Multiplying this by the column area, we get an ASD available compressive strength of 20.6 x 9.13 = 188.1 k—virtually the same as obtained by the calculations and Table 4-1a.

CASE STUDY—DESIGN IN STEEL (LRFD)

6.2.LRFD BEAM 3

For the design of typical members in LRFD, we'll follow steps similar to ASD.

Determine Loads (Beam 3—LRFD)

Calculate the uniform load imposed on Beam 3.

The load tributary area for Beam 3 is a section of floor 8 ft × 24 ft (Figure 6.5L). The uniformly distributed loads (w) on Beam 3 are:

$$\begin{aligned} w_{DL} &= \text{uniform dead load} \\ &= 75\ \text{lb/ft}^2 \times 8\ \text{ft} = 600\ \text{lb/ft} = 0.60\ \text{k/ft} \\ w_{LL} &= \text{uniform live load} \\ &= 75\ \text{lb/ft}^2 \times 8\ \text{ft} = 600\ \text{lb/ft} = 0.60\ \text{k/ft} \\ \hline w_{TL} &= \text{uniform total (service) load} = 1.20\ \text{k/ft} \end{aligned}$$

Applying the governing LRFD load combination (1.2 D + 1.6 L) to $w_{DL} + w_{LL}$ (see Chapter 4):

Total LRFD Design Load (w_u)

$$\begin{aligned} \mathbf{w_u} &= 1.2\ w_{DL} + 1.6\ w_{LL} \\ &= (1.2 \times 0.60) + (1.6 \times 0.60) = 0.72 + 0.96 \\ &= \mathbf{1.68\ k/ft} \end{aligned}$$

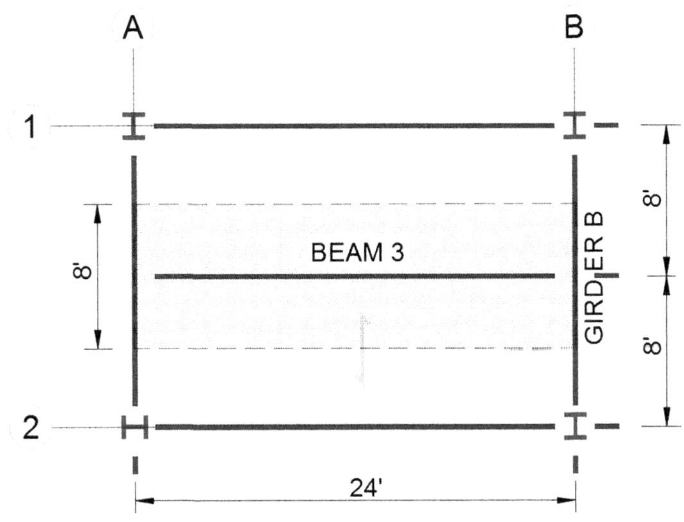

Figure 6.5L Load Tributary Area of Beam 3

Determine Reactions, Maximum Shear, Maximum Moment (Beam 3—LRFD)

Beam 3 is simply supported with a uniformly distributed load along its length. From Appendix 4—Beam Diagrams and Formulae—the shear and moment diagrams with maximum values for Girder B are shown in Figure 6.6L.

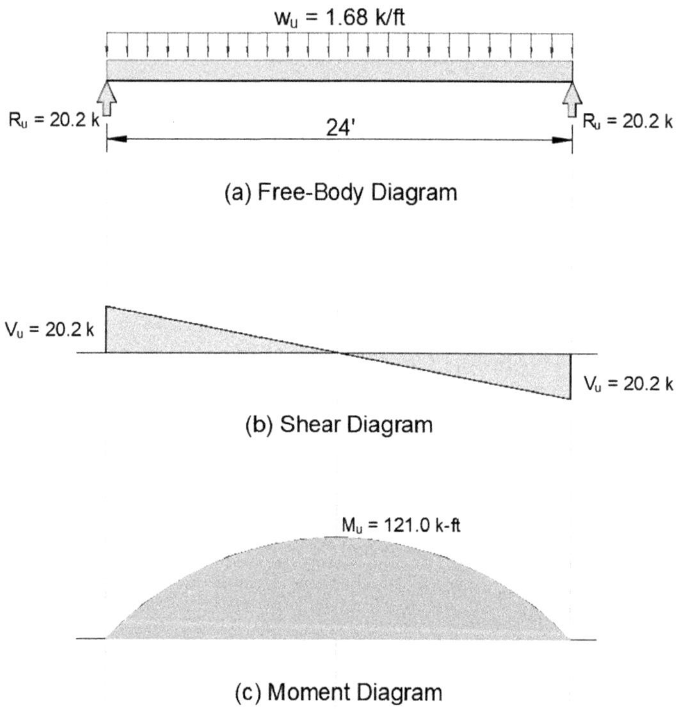

Figure 6.6L Beam 3: Free-body, Shear & Moment Diagrams

$L\ (\text{length}) = 24\ \text{ft}$

$w_u = 1.68\ \text{k/ft}$

Reactions (R_u)

$R_u = w_u L/2$

$R_u = (1.68 \times 24)/2$

$\mathbf{R_u = 20.2\ k}$

Maximum Shear (V_u)

$V_u = R_u$

$\mathbf{V_u = 20.2\ k}$

(Required Shear Strength)

Maximum Moment (M_u)

$M_u = w_u L^2/8$

$M_u = (1.68 \times 24^2)/8$

$\mathbf{M_u = 121.0\ k\text{-}ft}$

(Required Flexural Strength)

Design of Beam 3 (LRFD)

Step 1: Resisting Moment (Beam 3—LRFD)

Designing to resist moment involves selecting a beam with an available flexural strength equal to or greater than the required flexural strength, as expressed by:

$\mathbf{M_u \leq \Phi_b\, M_{px}}$

where:

M_u = required flexural strength

$\Phi_b\, M_{px}$ = available flexural strength

Calculate the Required Section Modulus (Z_x)

$$\mathbf{Z_x = M_u / (\Phi_b \times F_y)}$$
$$= (121.0 \times 12)/(0.9 \times 50)$$
$$Z_x = 32.3 \text{ in}^3$$

where:
Z_x = required section modulus (in^3)
M_u = maximum moment = 121.0 k-ft
Φ_b = strength reduction factor = 0.9
F_y = yield stress = 50 k/in^2

(for unit consistency, M_u is multiplied by 12 to convert ft to inches)

Select the Beam from Tables

From AISC Table 3-2 (W-Shapes—Selection by Z_x), select a shape with an available Z_x equal to or greater than that required.

A W14×22 shape with an available Z_x of 33.2 in^3 is the closest (but still greater) value to the required Z_x of 32.3 in^3. Also note that this shape's available flexural strength of 125 k-ft, is greater than the required flexural strength of 121.0 k-ft.

Shape	Available Section Modulus (Z_x)	Required Section Modulus (Z_x)
W14×22	33.2 in^3	32.3 in^3

Available Flexural Strength ($\Phi_b M_{px}$)	Required Flexural Strength (M_u)
125.0 k-ft	121.0 k-ft

We'll tentatively select the W14×22, (with an I_x of 199 in^4) for Beam 3 pending a shear and deflection check.

Step 2: Checking for Shear (Beam 3—LRFD)

Checking for shear involves assuring that its available shear strength is equal to or greater than its required shear strength, as expressed by:

$$\mathbf{V_u \le \Phi_v V_{nx}}$$

where:
V_u = required shear strength
$\Phi_v V_{nx}$ = available shear strength

From AISC Table 3-2, the available shear strength of a W14×22 is 94.5 k.

From the shear diagram, the required shear strength is 20.2 k.

Shape	Available Shear Strength ($\Phi_v V_{nx}$)	Required Shear Strength (V_u)
W14×22	94.5 k	20.2 k

Since the available shear strength is considerably greater than the required shear strength, the W14×22 passes the check for shear.

Step 3: Checking for Deflection (Beam 3—LRFD)

For the 24 ft long W14×22 beam, assure that its maximum deflection is equal to or less than its allowable deflection for both:

- live load (alone)
- total load (dead + live)

For LRFD, the code requires deflections to be checked with service loads. The service loads are:

$$w_{DL} = \text{uniform dead load (service)} = 0.60\ \text{k/ft}$$
$$w_{LL} = \text{uniform live load (service)} = 0.60\ \text{k/ft}$$
$$w_{TL} = \text{total uniform load (service)} = 1.20\ \text{k/ft}$$

Live Load Deflection

- Live Load: Maximum Deflection for the W14×22 Beam

$$\boldsymbol{\delta_{LL} = 5(w_{LL}L \times L^3) / 384\ EI}$$

where:
δ_{LL} = live load maximum deflection (in)
w_{LL} = uniform live load = 0.60 k/ft
L = length = 24 ft
E = modulus of elasticity = 29,000 k/in^2
I_x = moment of inertia = 199 in^4

$$\delta_{LL} = \left[(5) \times (0.6 \times 24) \times (24 \times 12)^3\right] / (384 \times 29{,}000 \times 199) = 0.78\ \text{in}$$

- Live Load: Allowable Deflection for the W14×22 Beam

$$\boldsymbol{\Delta_{LL} = L / 360}$$

where:
Δ_{LL} = live load allowable deflection (in)

$$\Delta_{LL} = (24 \times 12)/360 = 0.80\ \text{in}$$

Shape	Live Load Maximum Deflection (δ_{LL})	Live Load Allowable Deflection (Δ_{LL})
W14×22	0.78 in	0.80 in

Since the live load maximum deflection is less than the live load allowable deflection, the W14×22 passes the check for live load deflection.

Total Load Deflection

- Total Load: Maximum Deflection for the W14×22 Beam

$\boldsymbol{\delta_{TL} = 5\left(w_{TL}L \times L^3\right) / 384\ EI}$

where:

δ_{TL} = total load maximum deflection (in)
w_{TL} = uniform total load = 1.20 k/ft
L = length = 24 ft
E = modulus of elasticity = 29,000 k/in^2
I = moment of inertia = 199 in^4

$$\delta_{TL} = [(5) \times (1.2 \times 24) \times (24 \times 12)^3] / (384 \times 29{,}000 \times 199) = 1.55 \text{ in}$$

- Total Load: Allowable Deflection for the W14×22 Beam

$\boldsymbol{\Delta_{TL} = L / 240}$

where:

Δ_{TL} = total load allowable deflection (in)

$$\Delta_{TL} = (24 \times 12)/240 = 1.20 \text{ in}$$

Shape	Total Load Maximum Deflection (δ_{TL})	Total Load Allowable Deflection (Δ_{TL})
W14×22	1.55 in	1.20 in

Since the total load maximum deflection is greater than the total load allowable deflection, the W14×22 does NOT pass the check for total deflection.

Return to Table 3-2 and select the next larger economical shape W14×26 with an I_x of 245 in^4. We'll find that this also will not meet the deflection criteria and, as before, we'll return to the table and select the W16×26 with an I_x of 301 in^4.

- Total Load: Maximum Deflection for the W16×26 Beam

$$\delta_{TL} = \left[(5) \times (1.2 \times 24) \times (24 \times 12)^3\right] / (384 \times 29{,}000 \times 301) = 1.02 \text{ in}$$

Shape	Total Load Maximum Deflection (δ_{TL})	Total Load Allowable Deflection (Δ_{TL})
W16×26	1.02 in	1.20 in

Since the total load maximum deflection is less than the total load allowable deflection, the W16×26 passes the check for total load deflection, so we'll confirm it as our selection for Beam 3.

6.3.LRFD GIRDER B

For the design of Girder B, we'll follow steps similar to those in Beam 3.

Determine Loads (Girder B—LRFD)

Girder B is supporting Beam 3 and Beam 4 at its midspan. The combined design point load (P) from these two identical beams is 40.4 k (i.e., the end reactions of each beam; 20.2 k + 20.2 k) (Figure 6.7L).

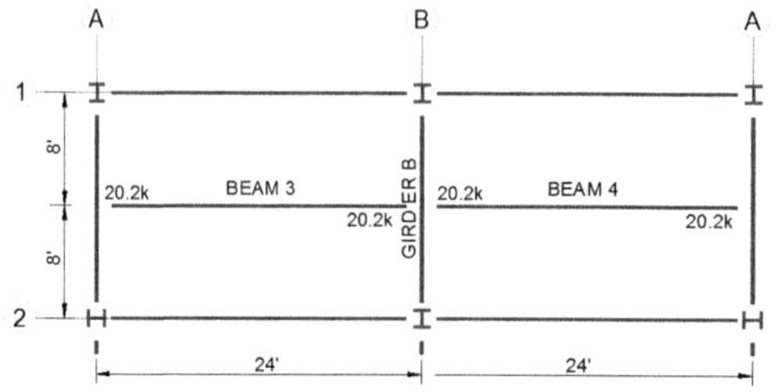

Figure 6.7L Loads on Girder B

Determine Reactions, Maximum Shear, Maximum Moment (Girder B—LRFD)

Girder B is simply supported with a point load at its midspan. From Appendix 4—Beam Diagrams and Formulae—the shear and moment diagrams with maximum values for Girder B are shown in Figure 6.8L.

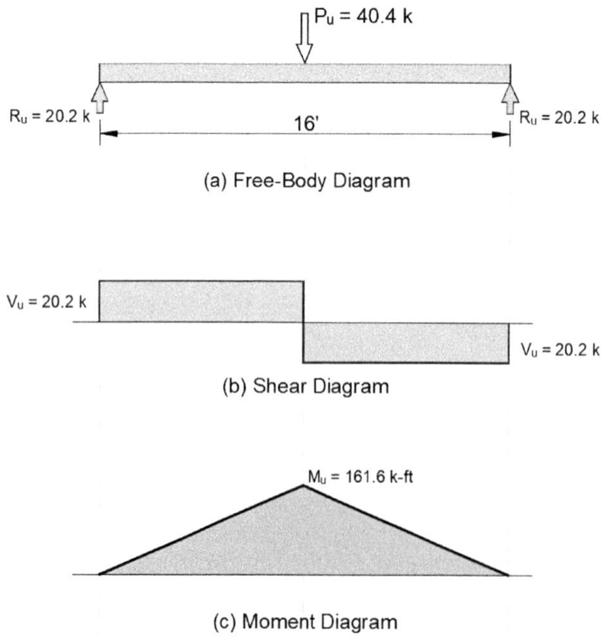

Figure 6.8L Girder B: Free-body, Shear & Moment Diagrams

$L\ (\text{length}) = 16\ \text{ft}$

$P_u = 40.4\ \text{k}$

Reactions (R_u)

$R_u = P/2$

$R_u = 40.4/2$

$\mathbf{R_u = 20.2\ k}$

Maximum Shear (V_u)

$V_u = R_u$

$\mathbf{V_u = 20.2\ k}$

(Required Shear Strength)

Maximum Moment (M_u)

$M_u = PL/4$

$M_u = (40.4 \times 16)/4$

$\mathbf{M_u = 161.6\ k\text{-}ft}$

(Required Flexural Strength)

Design of Girder B (LRFD)

Step 1: Resisting Moment (Girder B—LRFD)

Designing to resist moment involves selecting a beam with an available flexural strength equal to or greater than the required flexural strength, as expressed by:

$\mathbf{M_u < \Phi_b M_{px}}$

where:
M_u = required flexural strength
$\Phi_b\ M_{px}$ = available flexural strength

Calculate the Required Section Modulus (Z_x)

$$\mathbf{Z_x} = \mathbf{M_u} / (\mathbf{\Phi_b} \times \mathbf{F_y})$$
$$= (161.6 \times 12) / (0.9 \times 50)$$
(changing k-ft to k-in)
$$Z_x = 43.1 \text{ in}^3$$

where:
Z_x = required section modulus (in^3)
M_u = maximum moment = 161.6 k-ft
Φ_b = strength reduction factor = 0.9
F_y = yield stress = 50 k/in^2

Select the Girder from Tables

From AISC Table 3-2, select a shape with an available Z_x equal to or greater than that required.

A W16×26 shape with an available Z_x of 44.2 in^3 is the closest (but still greater) value to the required Z_x of 43.1 in^3. Also note that the W16×26 available flexural strength of 166 k-ft is greater than the required flexural strength of 161.6 k-ft.

Shape	Available Section Modulus (Z_x)	Required Section Modulus (Z_x)	Available Flexural Strength ($\Phi_b\ M_{px}$)	Required Flexural Strength (M_u)
W16x26	44.2 in^3	43.1 in^3	166 k-ft	161.6 k-ft

We'll tentatively select the W16×26, with an I_x of 301 in^4, for Girder B pending a shear and deflection check.

Step 2: Checking for Shear (Girder B—LRFD)

For the W16×26 girder, assure that its available shear strength is equal to or greater than its required shear strength, as expressed by:

$\mathbf{V_u \le \Phi_v V_{nx}}$

where:
V_u = required shear strength
$\Phi_v\ V_{nx}$ = available shear strength

From AISC Table 3-2 the available shear strength of a W16×26 is 106.0 k.
From the shear diagram, the required shear strength is 20.2 k.

Shape	Available Shear Strength ($\Phi_v V_{nx}$)	Required Shear Strength (V_u)
W16×26	106.0 k	20.2 k

Since the available shear strength is considerably greater than the required shear strength, the W16×26 passes the check for shear.

Step 3: Checking for Deflection (Girder B—LRFD)

For the 16 ft long W16×26 girder, assure that its actual deflection is equal to or less than its allowable deflection for both:

- live load (alone)
- total load (dead + live)

Recall from Beam 3 that the code requires LRFD deflections to be checked with service (i.e., un-factored) loads. The service loads are:

$$
\begin{aligned}
P_{DL} &= \text{point dead load (service)} = 14.4\ \text{k} \\
P_{LL} &= \text{point live load (service)} \quad = 14.4\ \text{k} \\
\hline
P_{TL} &= \text{total point service load} \quad = 28.8\ \text{k}
\end{aligned}
$$

Live Load Deflection

- Live Load: Maximum Deflection for the W16×26 girder:

$$\boldsymbol{\delta_{LL} = P_{LL}\, L^3/48\ EI}$$

where:

δ_{LL} = live load maximum deflection (in)

P_{LL} = point live load = 14.4 k

(The total point load is 28.8 k. Since the live load and dead loads are equal, the live point load (P) is ½, or 14.4 k.)

L = length = 16 ft

E = modulus of elasticity = 29,000 k/in²

I_x = moment of inertia = 301 in⁴

$$= \left[(14.4)\times(16 \text{ x } 12)^3\right]/(48\times 29{,}000\times 301) = 0.24 \text{ in}$$

- Live Load: Allowable Deflection for the W16×26 girder

$$\boldsymbol{\Delta_{LL} = L/360}$$

$$=(16 \text{ x } 12)/360 = 0.53 \text{ in}$$

where:

Δ_{LL} = live load allowable deflection (in)

Shape	Live Load Maximum Deflection (δ_{LL})	Live Load Allowable Deflection (Δ_{LL})
W16×26	0.24 in	0.53 in

Since the live load maximum deflection is less than the live load allowable deflection, the W16×26 passes the check for live load deflection.

Total Load Deflection

- Total Load: Maximum Deflection for the W16×26 girder:

$\delta_{TL} = P_{TL}\, L^3 / 48\, EI$

where:
δ_{TL} = total load maximum deflection (in)
P_{TL}= point total load = 28.8 k

$$= [(28.8) \times (16 \times 12)^3]/(48 \times 29{,}000 \times 301 = 0.49 \text{ in}$$

- Total Load: Allowable Deflection for the W16×26 girder

$$\Delta_{TL} = \mathbf{L / 240}$$
$$= (16 \times 12) / 240 = 0.80 \text{ in}$$

where:
Δ_{TL} = live load allowable deflection (in)

Shape	Total Load Maximum Deflection (δ_{TL})	Total Load Allowable Deflection (Δ_{TL})
W16×26	0.49 in	0.80 in

Since the total load maximum deflection is less than the total load allowable deflection, the W16×26 passes the check for total load deflection, so we'll confirm it as our selection for Girder B.

6.4.LRFD COLUMN B2

Determine Loads (Column B2—LRFD)

First Floor Loads

Column B2 is supporting first floor Girders B and E and Beams 5 and 6. The combined load from the end reaction of each of these members is tabulated below and shown in Figure 6.9L.

Member	First Floor Load on Column B2
Girder B	20.20 k
Girder E	20.20 k
Beam 5	20.20 k
Beam 6	20.20 k
Total	80.80 k

Roof Loads

In addition to the first floor loads, Column B2 is also supporting roof loads. Since the roof framing and loading is symmetrical about Column B2, we can calculate its roof load based on its load tributary area (Figure 6.10L).

Load tributary area for Column B2 $= 16\ \text{ft} \times 24\ \text{ft}$

$= 384\ \text{sf}$

D = uniform roof dead load = 30 psf

L_r = uniform roof live load = 30 psf

Applying the governing LRFD roof load combination:

$1.2\,D + 1.6\,L_r = (1.2 \times 30) + (1.6 \times 30) = 84\ \text{psf}$

Roof load on Column B2 $= 384\text{ft}^2 \times 84\ \text{lb/ft}^2$

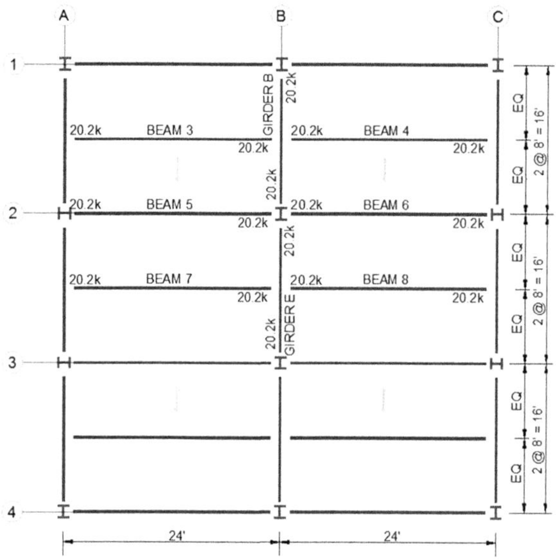

Figure 6.9L First Floor Loads on Column B2

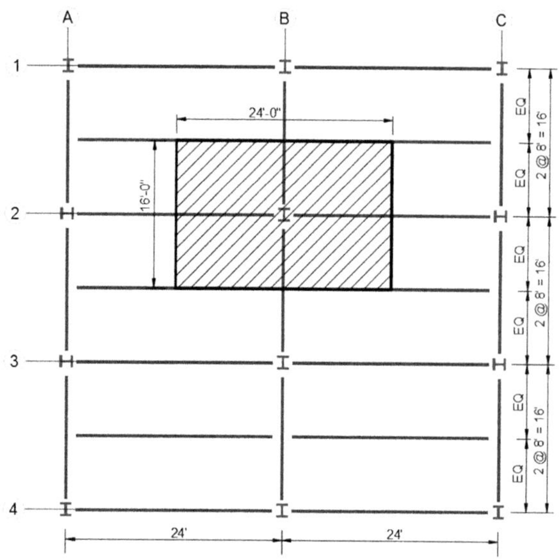

Figure 6.10L Roof Tributary Area for Column B2

Total Floor and Roof Loads on Column B2

First floor load	= 80.80 k
Roof load	= 32.30 k
Total Load on Column B2	= **113.10 k** (required compressive strength, P_u)

Establish the Effective Length (L_e) (Column B2—LRFD)

$L_e = kL = 12\ \text{ft}$

DESIGN OF COLUMN B2 (LRFD)

Designing to resist compression involves selecting a column with an available compressive strength equal to or greater than the required compressive strength—as expressed by the strength equation:

$\mathbf{P_u \leq \Phi_c P_n}$ where:

P_u = required compressive strength (i.e., loads)

$\Phi_c P_n$ = available compressive strength

Selecting a W Shape From Tables (Column B2—LRFD)

To select a W shape, we'll use AISC Table 4-1a (Available Strength in Axial Compression, kips—for W8 Shapes). This table provides the available strength of various W shapes for a given effective length (L_e) with respect to the shape's least radius of gyration (r), taking into account the column's slenderness ratio and column classification.

For an effective length (L_e) of 12 ft, enter Table 4-1a and select a W8×31 with an available LRFD compressive strength ($\Phi_c P_n$) of 283 k. This is the lightest W shape available in AISC column design tables for a 12 ft effective length.

Shape	Weight	Available Compressive Strength ($\Phi_c P_n$)	Required Compressive Strength (P_u)
W8×31	31 lb/ft	283 k	113.1 k

We see that the 283 k available compressive strength is significantly greater than the 113.1 k required compressive strength.

Selecting a Square HSS Shape from Tables (Column B2—LRFD)

As in ASD, using Table 4-4, we'll evaluate several sizes of square HSS sections with an effective length (L_e) of 12 ft:

Shape	Weight (lb/ft)	Available Compressive Strength ($\Phi_c P_n$)	Required Compressive Strength (P_u)
HSS 4½ × 4½ × 5/16	17.0	125.0 k	113.1 k
HSS 5 × 5 × 1/4	15.6	129.0 k	113.1 k
HSS 5½ × 5½ × 3/16	13.3	118.0 k	113.1 k
HSS 6 × 6 × 3/16	14.5	137.0 k	113.1 k

We see that the HSS 6 × 6 × 3/16 is the lightest section with an available compressive strength greater than the required compressive strength.

Discussion

Comparing our two column selections, we see that the weight of the HSS shape is significantly lower than the W shape.

ASD/LRFD DISCUSSION

Why Two Design Methodologies for Steel?

ASD and LRFD are two distinct and separate design methodologies, equally accepted by the AISC. The Reader may reasonably question why the AISC provides two methodologies for designing in steel. The reason lies partly with the historical development of the methodologies described in Chapter 2, and partly with the practicing professional's comfort level adapting to new approaches.

AISC Guidelines for Choice of Methodologies

The AISC offers the following *general* guidelines for the choice of methodologies:

For a live load to dead load ratio of less than 3 (e.g., 50 psf LL : 25 psf DL)

- LRFD is more economical, ASD is more conservative.

For a live load to dead load ratio greater than 3 (e.g., 100 psf LL : 25 psf DL)

- LRFD is more conservative, ASD is more economical.

For a live load to dead load ratio equal to 3 (e.g., 75 psf LL : 25 psf DL)

- LRFD and ASD will generally produce similar results.

Case Study Results

Let's examine the results from our ASD/LRFD Case Study. Recall that our Case Study live load to dead ratio equals 1 (75 psf LL : 75 psf DL)—i.e., the ratio is less than 3.

- For Beam 3, both ASD and LRFD resulted in a W16×26 shape—in other words, both methodologies produced the same result.
- For Girder B, ASD resulted in a W14x30 shape while LRFD resulted in a W16×26 shape—in other words, LRFD produced an approximate 13% economy over ASD in the weight of steel.
- For Column B2, both methodologies produced the same results. However, this comparison may not be an appropriate reflection since the column capacities are significantly greater than the light loads that were used.

Looking at Beam 3 and Girder B together, consistent with the AISC general guidelines, we see that LRFD was indeed more economical. Magnified over the entirety of a building, this economy can be significant.

Conclusions

Since most building live load to dead load ratios are less than 3, LRFD generally provides greater economy for designing in steel, and is the prevalent steel methodology in use today. As always, however, the final decision on choice of methodology is up to the designer based upon his/her judgment and preference.

7 Understanding Wood

7.1 SAWN LUMBER—MANUFACTURE AND PROPERTIES

Wood utilized for structural purposes is generally referred to as lumber. Lumber may be cut and used directly from felled trees (logs), or may be manufactured in various ways from wood pieces and components. Lumber cut from natural wood is referred to as *sawn*; lumber that is manufactured is referred to as *engineered* (Figures 7.1 and 7.2).

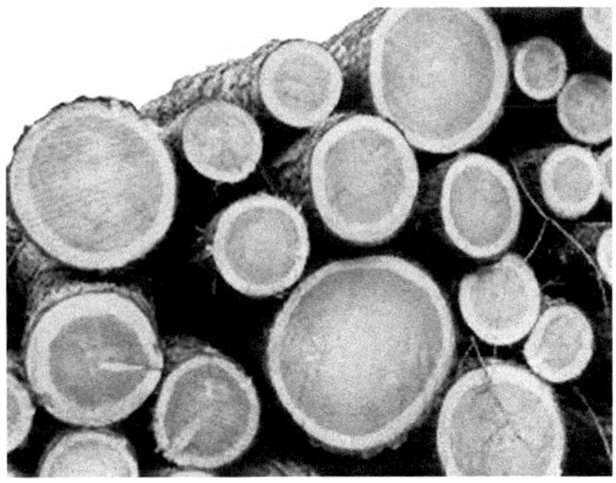

Figure 7.1 Wood Logging

Figure 7.2 A Wood Frame Building

Sawn lumber is commonly referred to by its nominal (rough cut) size such as 2×6, 4×6, 6×6, etc. Although terminology varies, members nominally 4 inches or less in the least dimension are generally referred to as dimensional lumber (Figure 7.3), while members 5 inches or greater in the least dimension are generally referred to as timbers (Figure 7.4). When lumber is seasoned and surfaced, its actual dimensions (referred to as dressed sizes) are slightly less than the nominal. For example, a nominal 2×6 will actually be 1½″×5½″ after being dressed. Actual dimensions must be used when designing. S4S (surfaced 4 sides) is the term for dressed sawn lumber.

Figure 7.3 Dimensional Lumber

Figure 7.4 Wood Timbers

Species and Grading

Sawn lumber is *anisotropic*, meaning that its structural characteristics vary depending upon the orientation of the grain.

The strength of sawn lumber is determined by its species, grading for quality, and grain orientation. Properties for various species, grades, and grain orientations are given in the National Design Specification for Wood Construction (NDS).

There are several accredited grading rules agencies in the US and Canada that define grading standards and reference design values (also called reference stresses). These agencies are referred to by various acronyms in the NDS tables. For the Case Study we'll use the design values provided by the West Coast Lumber Inspection Bureau (WCLIB).

Structural lumber is generally graded in descending quality as Select Structural, No. 1, No. 2, No. 3, Stud, Construction, Standard, and Utility. Further variations in reference stresses for the same species and grade may yet exist for members with larger cross-sectional sizes (Figure 7.5).

Figure 7.5 Wood Grading

Cellular Structure and Grain Orientation

In a cross section of a tree trunk (Figure 7.6), the *pith* is at the very center. Surrounding the pith is the *heartwood* containing dead cells that no longer serve a purpose except to support the tree. Next is the *sapwood* (often lighter in color than the heartwood) that carries water and nutrients between roots and leaves. Outside the sapwood close to the surface is the *cambium*, a thin layer of living cells that manufacture wood as they grow. These cells grow outward each spring, forming rings known as *annular growth rings* while producing two kinds of cells—longitudinal cells that align with the axis of the trunk, and ray cells that align perpendicular to the axis and extend outwards in cross section.

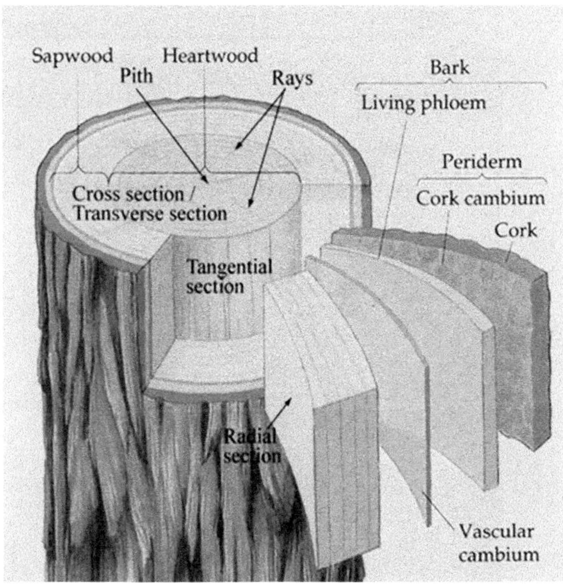

Figure 7.6 Cross Section of a Tree Trunk

The alignment of the longitudinal cells is what is referred to as direction of grain (Figure 7.7)—therefore the direction of grain is generally considered parallel to the length of a wood member (Figure 7.8).

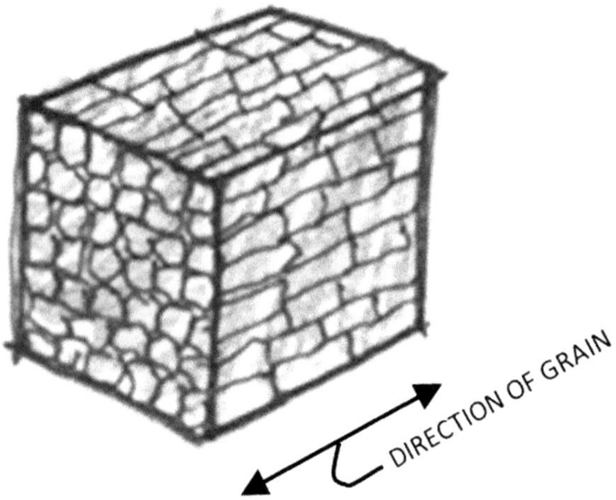

Figure 7.7 Alignment of Logintudinal Cells in Wood

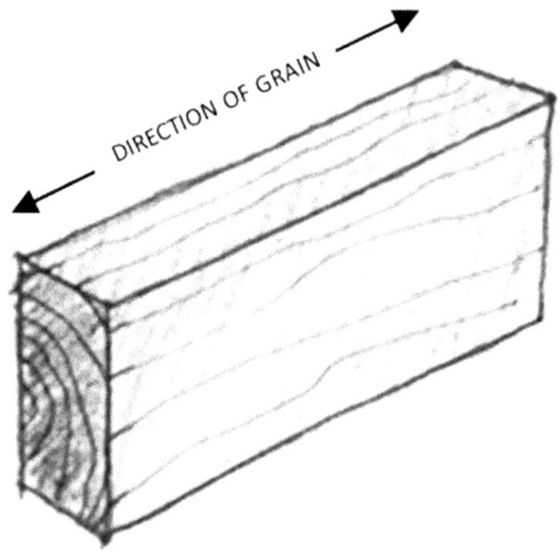

Figure 7.8 Direction of Grain in a Typical Wood Member

Stresses and Grain Orientation

Since grain orientation affects the structural properties of sawn wood, several types of stresses must be considered (Figure 7.9).

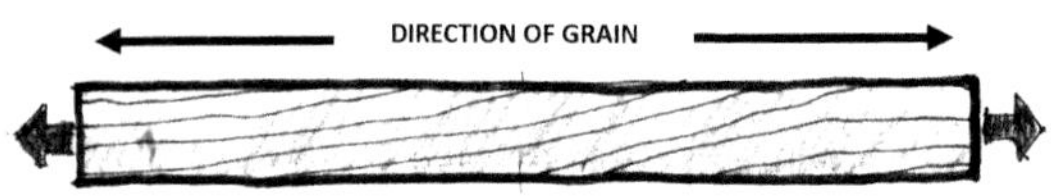

(a) Tensile Stress Parallel to Grain (F_t)

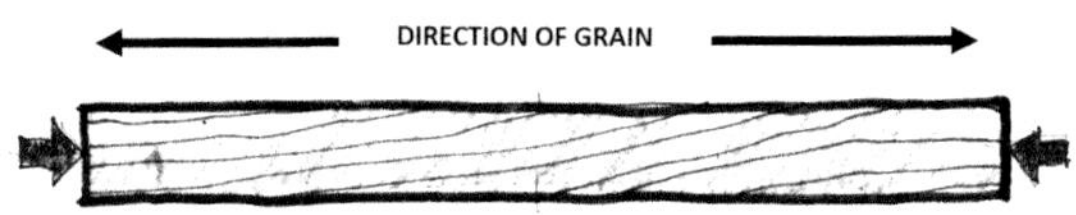

(b) Compressive Stress Parallel to Grain (F_c)

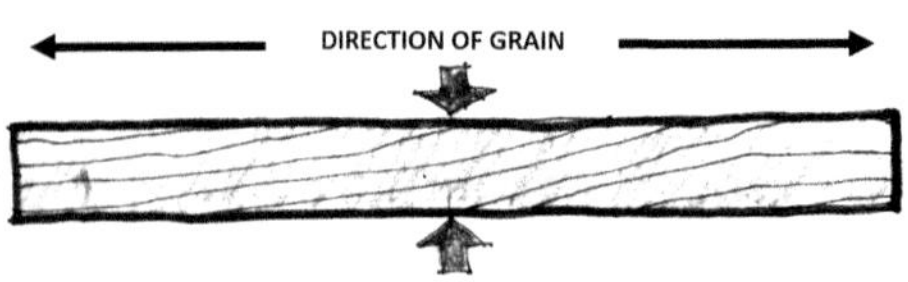

(c) Compressive Stress Perpendicular to Grain ($F_{c\perp}$)

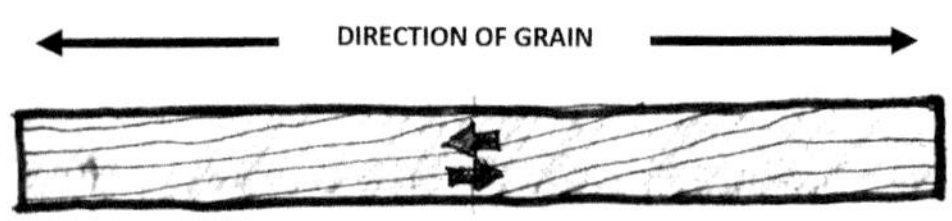

(d) Shear Stress Parallel to Grain (F_v)

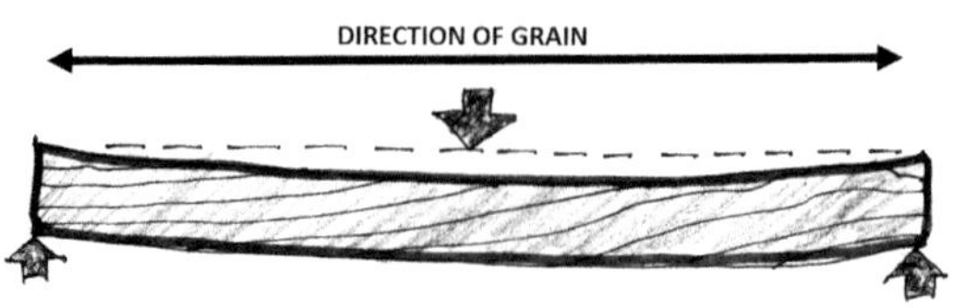

(e) Bending Stress (Fb)

Figure 7.9 Stresses and Grain Orientation

Reference Design Values

Reference design values are baseline stresses for various grain orientations that take safety factors into account. Reference design values depend on grading and whether the member is considered dimensional lumber or timber. NDS Table 4A (Reference Design Values for Visually Graded Dimension Lumber) and Table 4D (Reference Design Values for Visually Graded Timbers) provide reference design values for various species and grades (see Appendix 2).

Adjustment Factors and Adjusted Design Values

The NDS lists 14 *adjustment factors* that account for variations in conditions of use. Adjustment factors are based on the design methodology being used (i.e., Allowable Stress Design or Load and Resistance Factor Design), and are applied to reference design values, resulting in *adjusted design values*. An adjusted design value can be thought of as an "*allowable stress*", and we will generally refer to it as such.

For wood, a reference design value is denoted by (F), an adjusted design value (i.e., an allowable stress) by (F′), and an actual stress by (f).

Adjusted Design Value (F′) (i.e., Allowable Stress) =
Reference Design Value (F) × Applicable Adjustment Factors

Applicable adjustment factors are specified by the NDS for various reference design values (such as F_b, F_v, F_c, etc.) The designer is responsible for evaluating the conditions of use and applying the appropriate values for adjustment factors according to given criteria in the NDS. The various NDS adjustment factors are listed in Table 7.1. Except as noted the adjustment factors apply to both Allowable Stress Design and LRFD.

Table 7.1 NDS Adjustment Factor

Factor	Symbol	Factor	Symbol
Load duration factor (Allowable Stress Design only)	C_D	Repetitive member factor	C_r
Wet service factor	C_M	Column stability factor	C_P
Temperature factor	C_t	Buckling stiffness factor	C_T
Beam stability factor	C_L	Bearing area factor	C_b
Size factor	C_F	Format conversion factor (LRFD only)	K_F
Flat use factor	C_{fu}	Resistance factor (LRFD only)	Φ
Incising factor	C_i	Time effect factor (LRFD only)	λ

The following is a brief description of the 11 adjustment factors used in Allowable Stress Design. An adjustment factor of less than 1 tends to decrease the allowable stress, while an adjustment factor of greater than 1 tends to increase the allowable stress. The Reader is referred to the NDS for additional information on adjustment factors.

Load Duration Factor (C_D)

Wood is good at absorbing short-term loads such as those for wind, seismic, snow, and construction. The load duration factor applied for loads such as these is greater than 1, thereby tending to increase the allowable stress. The load duration factor applied for longer-term loads that cause wood to creep and permanently deflect is less than 1.

Wet Service Factor (C_M)

Wood weakens when used in areas of high moisture. A wet service factor less than 1 is applied for such areas.

Temperature Factor (C_t)

Wood weakens when used in areas of high temperature. A temperature factor less than 1 is applied for such areas.

Beam Stability Factor (C_L)

A beam is not as effective when unbraced. A beam stability factor less than 1 is applied for such members.

Size Factor (C_F)

In a member having a large cross section, the probability of a defect in its interior is greater. A size factor less than 1 is applied for such members.

Flat Use Factor (C_{fu})

For a member used flat (as in a floor board), the member is considered more stable. A flat use factor greater than 1 is applied for such members.

Incising Factor (C_i)

When holes are made in a member during the pressure treating process, an incising factor less than 1 is applied.

Repetitive Member Factor (C_r)

When three or more members are ganged together, the chance of a defect in multiple members is reduced. For such situations, a repetitive factor greater than 1 is applied.

Column Stability Factor (C_p)

(See Section 7.3 Sawn Lumber—Design Considerations for Columns.)

Buckling Stiffness Factor (C_T)

This factor recognizes the contribution of plywood sheathing to the buckling resistance of compression truss chords under specified conditions.

Bearing Area Factor (C_b)

This factor is used to increase reference compressive design value for concentrated loads on wood perpendicular to grain.

Modulus of Elasticity

Wood's modulus of elasticity (E) varies by whether the member is classified as dimensional lumber or timber. Within these classifications, it varies with the species and grade.

Two reference design values of modulus of elasticity are listed in the NDS:

- E (when working with deflection)
- E_{min} (when working with lateral stability)

Values for E and E_{min} are given in NDS Reference Design Value tables. When multiplied by the applicable adjustment factors, they become:

- E′ (adjusted modulus of elasticity for deflection)
- E_{min}' (adjusted modulus of elasticity for lateral stability)

The applicable adjustment factors for E in Allowable Stress Design are: (C_m, C_t, C_i).

The applicable adjustment factors for E_{min} in Allowable Stress Design are: (C_m, C_t, C_i, C_T).

7.2 SAWN LUMBER—DESIGN CONSIDERATIONS FOR BEAMS

The NDS accepts both Allowable Stress Design and Load and Resistance Factor Design (LRFD) methodologies. We'll use Allowable Stress Design for the Case Study, to which the following discussion refers.

Resisting Moment in a Beam

Designing a wood beam to resist moment involves having the member's actual bending stress be less than or equal to the allowable bending stress.

$$\mathbf{f_b \leq F'_b}$$

where:
f_b = actual bending stress
F'_b = allowable bending stress

Determining the Allowable Bending Stress (F'_b)

To determine F'_b in Allowable Stress Design, the NDS requires that the reference bending design value (F_b) be multiplied by the specified adjustment factors below.

In Allowable Stress Design:

$$\mathbf{F'_b = F_b \times (C_M \times C_t \times C_L \times C_F \times C_{fu} \times C_r)}$$

where:
F'_b = allowable bending stress
F_b = reference bending design value
$C_{(M,\, t,\, L,\, F,\, fu,\, r)}$—see Table 7.1

The flexure formula ($F = M / S$) is used to determine the required section modulus of a typical beam or girder. For wood, this is expressed as:

$$F'_b = M / S$$

or

$$S = M / F'_b$$

where:
S = required section modulus
M = maximum moment

Resisting Horizontal Shear in a Beam

Beams subject to vertical loads are subject to both vertical and horizontal shear stresses (i.e., shear parallel to grain). Horizontal shear stress is of greater concern in sawn wood beams than in beams of other materials, since wood members are weaker in the direction parallel to grain, tending to cause longitudinal splitting.

Within the cross section of a beam, the maximum horizontal shear stress occurs at the neutral axis, and reduces towards the top and bottom. The magnitude of horizontal shear stress is directly proportional to that of the vertical shear force. For a simply supported beam under a uniformly distributed load, the horizontal shear stress is a maximum at the ends where the vertical shear force is greatest (Figure 7.10).

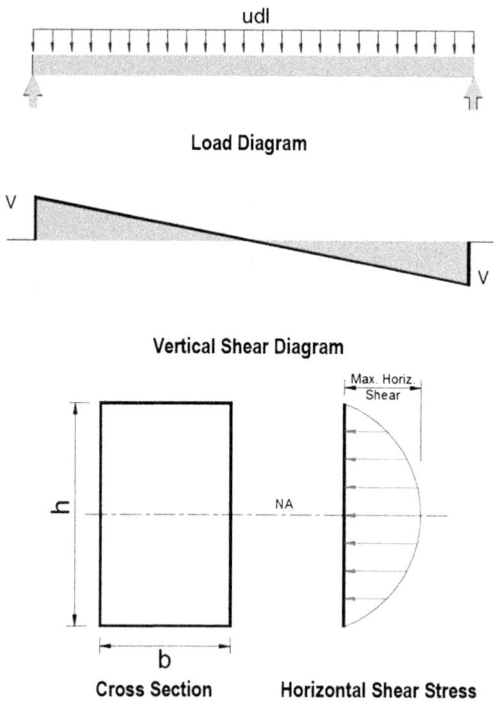

Figure 7.10 Horizontal Shear Stress in a Rectangular Beam

The maximum horizontal shear stress in a rectangular beam is given by:

$$f_v = 1.5 V / A$$

where:
f_v = maximum (actual) horizontal shear stress
V = maximum vertical shear force
A = area of cross section

Checking a wood beam for horizontal shear involves assuring that its actual horizontal shear stress is less than or equal to the allowable horizontal shear stress.

$$f_v \leq F'_v$$

where:
f_v = actual horizontal shear stress
F'_v = allowable horizontal shear stress

Determining the Allowable Horizontal Shear Stress (F'_v)

To determine F'_v in Allowable Stress Design, the NDS requires that the reference horizontal shear design value (F_v) be multiplied by the specified adjustment factors below.

In Allowable Stress Design:

$$\mathbf{F'_v = F_v \times (C_D \times C_M \times C_t \times C_i)}$$

where:
F'_v = allowable horizontal shear stress
F_v = reference horizontal shear design value
$C_{(D, M, t, i)}$ = applicable adjustment factors—see Table 7.1

7.3 SAWN LUMBER—DESIGN CONSIDERATIONS FOR COLUMNS

The NDS classifies wood columns into three categories:

1. solid wood columns
2. spaced columns
3. built-up columns

For the solid wood columns we'll use in the Case Study, the NDS recommends that the slenderness ratio (kL/d) shall not exceed 50 (except during construction where it shall not exceed 75).

The capacity of a wood column is given by:

$$\mathbf{P = F'_c \times A}$$

where:
P = column capacity
F'_c = allowable compressive stress parallel to grain
A = area of cross section

Determining the Allowable Compressive Stress (F'_c)

To determine F'_c in Allowable Stress Design, the NDS requires that reference compressive design value (F_c) be multiplied by the specified adjustment factors.

In Allowable Stress Design:

$$\mathbf{F'_c = F_c \times (C_D \times C_M \times C_t \times C_F \times C_i \times C_p)}$$

where:
F'_c = allowable compressive stress parallel to grain
F_c = reference compressive design value parallel to grain
$C_{(D,M,t,F,i,p)}$ = applicable adjustment factors—see Table 7.1

Consistent with the standard column curve, the above equation provides the allowable compressive stress for columns—higher for short columns, lower for longer columns.

The Column Stability Adjustment Factor (C_P)

The column stability adjustment factor (C_p) ensures that weak-axis buckling or torsional buckling does not occur over the column's laterally unsupported lengths. A C_p of 1 is used for continuously braced columns (i.e., where buckling is not a concern) and a lower C_p is used for unbraced columns. Since most columns are not continuously braced, a column stability adjustment factor of less than 1 essentially reduces the column's capacity.

C_p is given by:

$$C_p = \frac{1+\left(F_{cE}/F_c^*\right)}{2c} - \sqrt{\left[\frac{1+\left(F_{cE}/F_c^*\right)}{2c}\right]^2 - \frac{F_{cE}/F_c^*}{c}}$$

where:
C_p = column stability factor
F_{cE} = critical buckling design value
F_c^* = adjusted reference compressive design value parallel to grain
c = 0.8 for sawn lumber

F_{cE} is given by:

$$F_{cE} = \frac{0.822\ E_{min}'}{\left(\frac{l_e}{d}\right)^2}$$

where:
E_{min}' = adjusted modulus of elasticity
l_e = effective length
d = least dimension of column

F^*_c is given by:

$$\mathbf{F_c^* = F_c \times (C_D \times C_M \times C_t \times C_F \times C_i)}$$

where:
F_c = reference compressive design value parallel to grain
$C_{(D,\ M,\ t,\ F,\ i)}$ = applicable adjustment factors—see Table 7.1

7.4 ENGINEERED LUMBER—MANUFACTURE AND PRODUCTS

Since a sawn member's section dimensions and length are limited by tree size, the need for larger members has contributed to the relatively recent innovation and increased use of engineered wood.

Engineered wood generally refers to wood products that are manufactured by industrial processes. Such products are made by bonding together wood veneers, strands, or lumber, to produce a larger composite material with consistent properties. For greater economy, high-quality laminations are sometimes located in high stress areas (such as outer tension and compression zones) and lesser quality laminations near lower stress areas (near the neutral axis).

The origins of the engineered wood industry can be traced back to the 1860s when the first plywood panel was conceived. Plywood's development for the building industry languished however until the invention of waterproof adhesive in the 1930s, and the intensive use of plywood panels during World War II for quickly constructed military barracks. With the end of the war, the use of plywood quickly spread to the civilian market, leading to the rise of other engineered wood products and the entire engineered wood industry as it is today.

From a structural perspective, engineered wood products are treated as a homogeneous material with reliable design values provided by the manufacturer.

Figure 7.11 An Engineered Glulam Structure

Common engineered structural wood products are:

- Panels such as plywood and oriented strand board (OSB), usually 4 ft × 8 ft, are commonly used for subfloors and wall and roof sheathing. These panels typically provide the lateral resistance for wood-framed structures.
- Structural members such as I-joists, structural composite lumber (SCL), and glued laminated timber (glulam) (Figure 7.11).

Structural Products

Wood I-Joists

I-joists are commonly used in floor construction in light wood frame construction and are available in various depths and thicknesses. They typically consist of a plywood or OSB web attached to laminated veneer or sawn lumber flanges (Figure 7.12).

Figure 7.12 I-Joists

From a structural perspective, an I-joist uses material more efficiently than a solid sawn member by placing material farther away from the neutral axis at the flanges. The flanges primarily resist bending stresses, while the web primarily provides shear resistance.

Structural Composite Lumber (SCL)

Structural composite lumber refers to a family of engineered wood products produced by bonding wood veneers or strands into long blocks of material called billets. These are then cut to size in various sizes and thicknesses for use as structural members.

Types of SCL include laminated veneer lumber (LVL), laminated strand lumber (LSL), and parallel strand lumber (PSL). In general LVL and LSL are used for lighter members, PSL for heavier members (Figure 7.13).

Figure 7.13 Structural Composite Lumber (SCL)

Glued Laminated Timber (Glulam)

Glued laminated timber is made by bonding together individual layers of lumber, two inches thick or less, that can be formed into any size or shape, including curves. Glulam is an extremely versatile product that can be used for a variety of structural members including arches (Figure 7.14).

Figure 7.14 Glued Laminated Timber (Glulam)

Mass Timber

Mass timber refers to a construction method gaining in popularity and acceptance that uses thick, solid large prefabricated wood panels as structural elements such as floors and walls (Figure 7.15).

Mass timber construction is not yet a code-recognized type of construction, but its technology is rapidly evolving with new products and construction methods being created. Two common types of mass timber panels are cross laminated timber (CLT), and nail laminated timber (NLT).

Figure 7.15 Mass Timber Construction

Cross Laminated Timber (CLT)

- CLT panels are created from several layers of kiln-dried lumber boards stacked in alternating directions, bonded with structural adhesives (Figure 7.16).

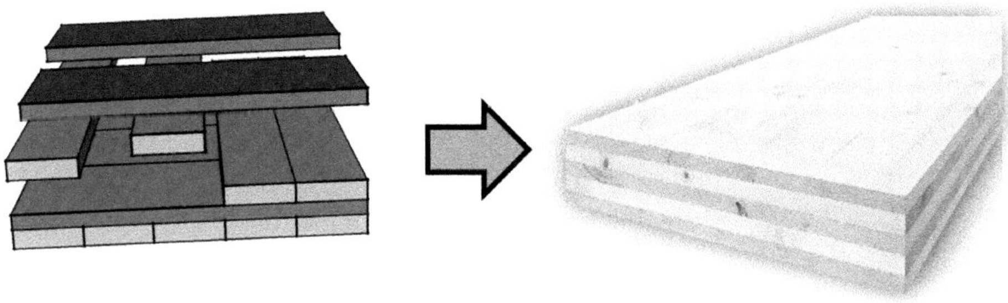

Figure 7.16 Cross Laminated Timber Panel

Nail Laminated Timber

- NLT panels are created from dimensional lumber (such as 2×4, 2×6, 2×8, etc.) stacked on edge and fastened together with nails (Figure 7.17).

Figure 7.17 Nail Laminated Timber Panel

7.5 ENGINEERED LUMBER—DESIGN CONSIDERATIONS

Manufacturers of engineered wood products provide design tables that simplify the selection of various members. These tables provide the moment, shear, deflection, and member self-weights for various loading and span conditions.

For the Case Study in wood, we'll first design using sawn lumber, then design using engineered lumber.

8

Design in Sawn Wood—Case Study

Our Case Study for design in sawn wood is a two-story one-way framed structure, 48 ft × 48 ft, with beams spanning 24 ft, girders spanning 16 ft, and 12 ft floor-to-floor heights. We'll focus on the structural design of the first floor and select typical members Beam 3, Girder B, and Column B2 to design. In addition to the loads from the first floor, the loads from the roof will be added onto Column B2. Figure 8.1 shows the first floor framing plan.

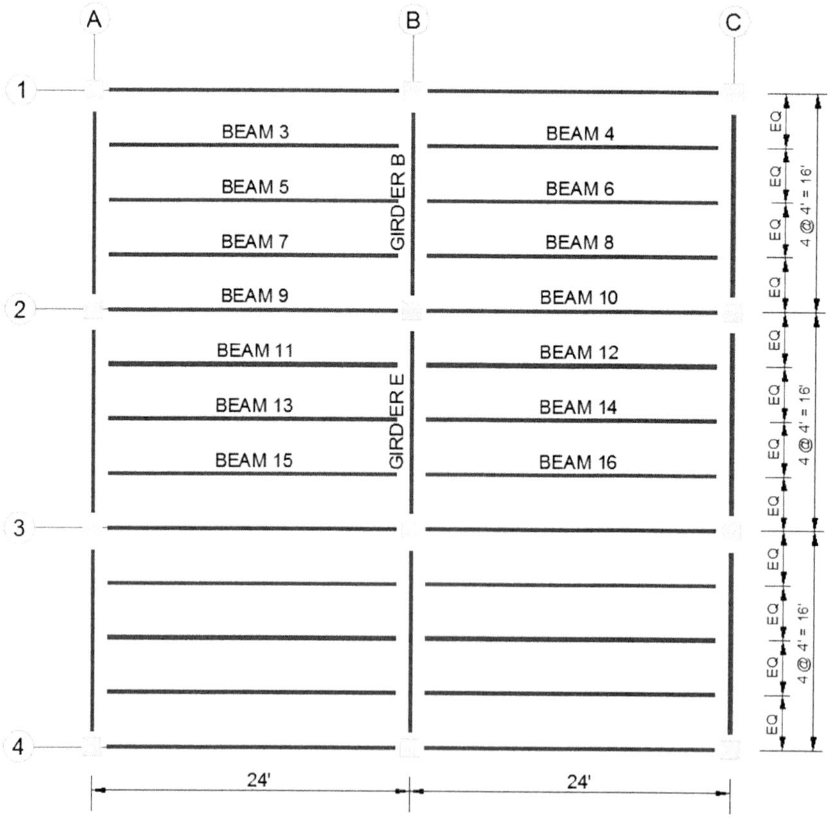

Figure 8.1 First Floor Framing Plan

We'll design the typical members using Allowable Stress Design based on the National Design Specification (NDS) Package for Wood Construction. This document contains tables from which we'll select structural members and perform checks (see Appendix 2).

8.1 ASSUMPTIONS

Floor Construction

Floor construction will be 3″ nominal thickness wood decking, supported on wood beams as a one-way spanning system (Figure 8.2).

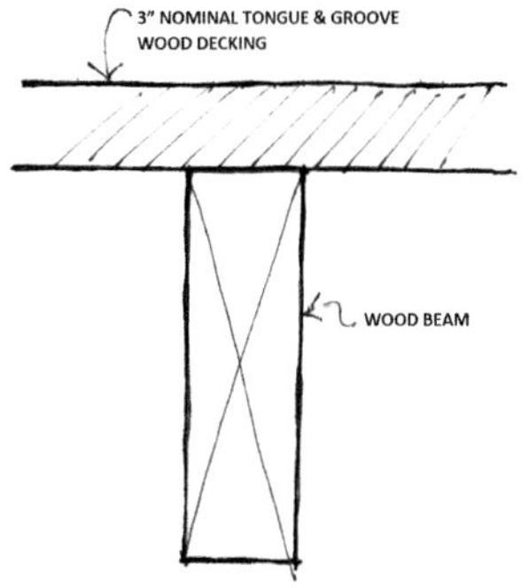

Figure 8.2 Floor Construction

Adjustment Factors

Except for the column stability factor (C_p), all applicable adjustment factors will be assumed to be 1.

Member Shapes and Material Properties

Structural members will be rectangular-shaped sawn lumber with the following relevant material properties:

- Species: Douglas Fir-Larch
- Commercial grade: Dense Select Structural

We'll assume our members' shapes to be 5″ × 5″ or larger. Table 4D in Appendix 2 (Reference Design Values for Visually Graded Timbers) provides reference design values and moduli of elasticity for wood members 5″ × 5″ and larger. From this table, the relevant member properties for the Case Study as given by WCLIB Grading Rules Agency are:

Reference Design Value		**Applicable Adjustment Factors (assumed to be 1)**	**Adjusted Design Value**
Bending	F_b = 1,900 psi	($C_M \times C_t \times C_L \times C_F \times C_{fu} \times C_r$)	F'_b = 1,900 psi (Allowable Stress)
Shear parallel to grain (horizontal shear)	F_v = 170 psi	($C_D \times C_M \times C_t \times C_i$)	F_v = 170 psi (Allowable Stress)
Compression parallel to grain	F_c = 1,350 psi	See Case Study	See Case Study
Modulus of elasticity (for deflection)	E = 1,700,000 psi	(C_m, C_t, C_i)	E′ = 1,700,000 psi
Modulus of elasticity (for lateral stability)	E_{min} = 620,000 psi	(C_m, C_t, C_i, C_T)	E_{min} = 620,000 psi

Floor and Roof Loads

We'll assume the following service loads:

Floor Loads		Roof Loads	
Dead load		***Dead load***	
*self-weight of beam, floor, and finishes	= 12 psf	*self-weight of roof framing and finishes	= 12 psf
mechanical (pipes, ducts, etc.)	= 8 psf	mechanical (pipes, ducts, etc.)	= 8 psf
Total	**= 20 psf**	**Total**	**= 20 psf**
Live load		***Live load***	
Total	**= 40 psf**	**Total**	**= 30 psf**

*In wood design, the self-weight of structural members is relatively small in relation to the other dead loads and is typically accounted for by including a suitable square foot allowance in the floor dead load calculations.

Beams and Girders

Beams and girders will be considered to have:

- simple supports
- full lateral bracing

Figure 8.3 shows the typical steps we'll follow in the design of Beam 3 and Girder B.

1. design to resist moment
2. check for shear
3. check for deflection

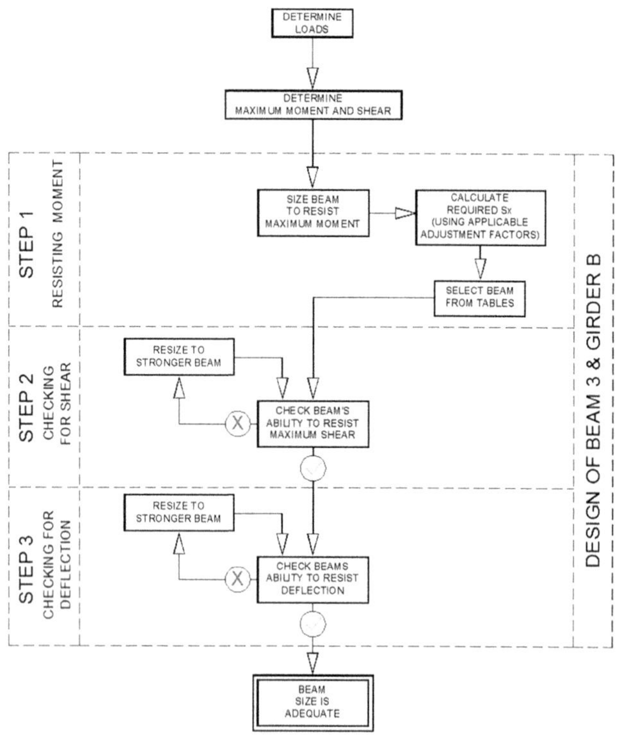

Figure 8.3 Flow Chart for Design of Beam 3 and Girder B

Columns

Columns will be considered to have:

- axial loads only, not subject to lateral loads
- floor-to-floor height (L) = 12 ft
- pinned top and bottom restraints; therefore, L_e k-factor = 1, and the effective length kL = 12 ft

Additional column assumptions will be given in the design of Column B2.

Figure 8.4 shows the typical steps we'll follow in the design of Column B2.

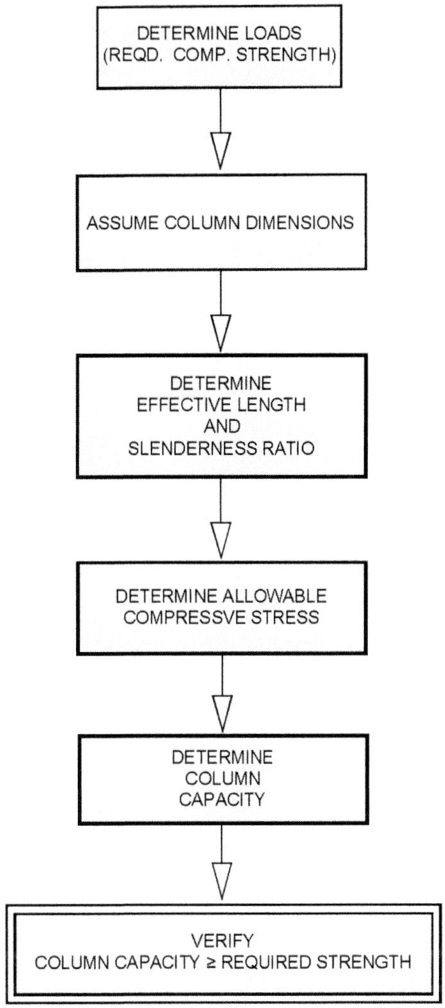

Figure 8.4 Flow Chart for Design of Column B2

Let's proceed to design Beam 3, Girder B, and Column B2, based on Allowable Stress Design.

8.2 BEAM 3

Determine Loads (Beam 3)

Calculate the uniform load imposed on Beam 3.

The load tributary area for Beam 3 is a section of floor 4 ft × 24 ft (Figure 8.5). The uniformly distributed loads (w) on Beam 3 are:

$$
\begin{aligned}
w_{DL} &= \text{uniform dead load} \\
&= 20\ \text{lb/ft}^2 \times 4\ \text{ft} = 80\ \text{lb/ft} \ \ = 0.08\ \text{k/ft} \\
w_{LL} &= \text{uniform live load} \\
&= 40\ \text{lb/ft}^2 \times 4\ \text{ft} = 160\ \text{lb/ft} = 0.16\ \text{k/ft} \\
\hline
w_{TL} &= \text{uniform total (service) load} = 0.24\ \text{k/ft}
\end{aligned}
$$

Applying the governing Allowable Stress Design load combination (1.0 D + 1.0 L) to $w_{DL} + w_{LL}$ (see Chapter 4):

Total Design Load (w)

$$
\begin{aligned}
\mathbf{w} &= 1.0\,w_{DL} + 1.0\,w_{LL} \\
&= (1.0 \times 0.08) + (1.0 \times 0.16) \\
&= \mathbf{0.24\ k/ft}
\end{aligned}
$$

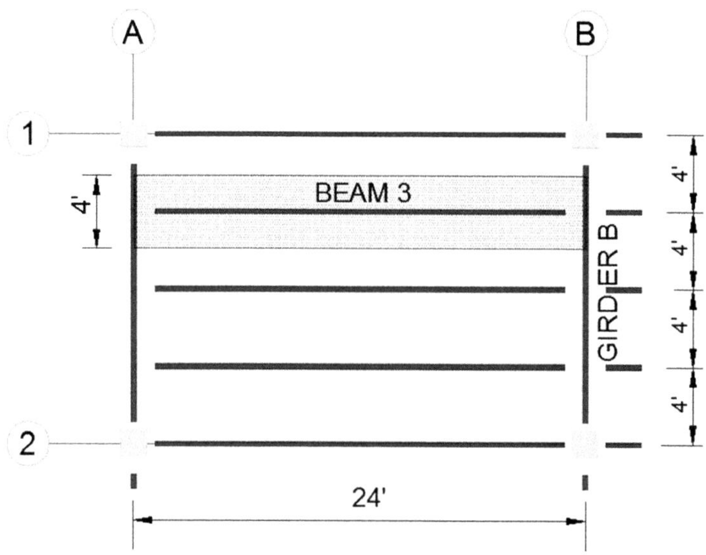

Figure 8.5 Load Tributary Area of Beam 3

Determine Reactions, Maximum Shear, Maximum Moment (Beam 3)

Beam 3 is simply supported with a uniformly distributed load along its length. From Appendix 4—Beam Diagrams and Formulae—the shear and moment diagrams with maximum values for Beam 3 are shown in Figure 8.6.

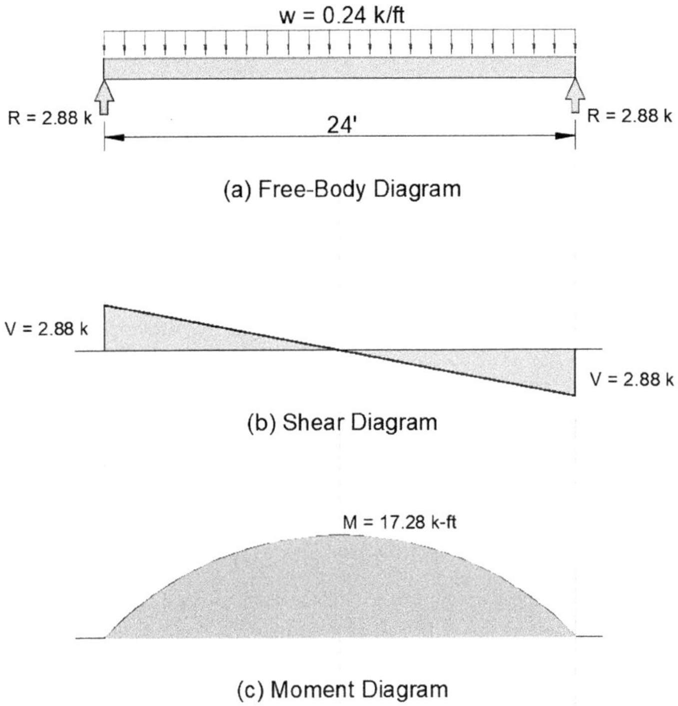

$L\ (\text{length}) = 24\ \text{ft}$

$w = 0.24\ \text{k/ft}$

Reactions

$R = wL/2$

$R = (0.24 \times 24)/2$

$\mathbf{R = 2.88\ k}$

Maximum Shear

$V = R$

$\mathbf{V = 2.88\ k}$

Maximum Moment

$M = wL^2/8$

$M = (0.24 \times 24^2)/8$

$\mathbf{M = 17.28\ k\text{-}ft}$

Figure 8.6 Beam 3: Free-body, Shear & Moment Diagrams

Design of Beam 3

Step 1: Resisting Moment (Beam 3)

Determine the Allowable Bending Stress (F'_b)

See Assumptions: 1,900 psi

Calculate the Required Section Modulus (S)

$$\mathbf{S = M / F'_b}$$
$$= (17.28 \times 12 \times 1{,}000) / 1{,}900$$
$$= 109.14\ \text{in}^3$$

where:
S = required section modulus (in^3)
M = maximum moment = 17.28 k-ft

(for unit consistency, M is multiplied by 12 to convert ft to inches, and by 1,000 to convert k to lbs)

Select the Beam from Tables

From Table 1B in Appendix 2 (Section Properties of Standard Dressed [S4S] Sawn Lumber), select a shape with an available S_{xx} greater than that required. Let's consider two viable options:

Nominal Beam Size (b × d)	S_{xx} (in^3)	Area (in^2)	Required Section Modulus (S)
6×12	121.2	63.25	109.14 in^3
4×16	135.6	53.38	

In the interest of minimizing floor-to-floor heights, we'll tentatively select the shallower 6×12 (with I_{xx} of 697.1 in^4 and cross-sectional area A of 63.25 in^2), pending a shear and deflection check.

For the 6×12, let's compare the actual bending stress to the allowable bending stress.

$$\begin{aligned} \mathbf{f_b} &= \mathbf{M / S} \\ &= (17.28 \times 12 \times 1{,}000)/121.2 \\ &= 1{,}710.9 \text{ psi} \end{aligned}$$

Actual Bending Stress (f_b)	Allowable Bending Stress (F'_b)
1,710.9 psi	**1,900 psi**

We see that the actual bending stress is less than the allowable.

Step 2: Checking for Horizontal Shear (Beam 3)

Checking the 6×12 wood beam for horizontal shear involves assuring that the maximum horizontal shear stress (f_v) is less than or equal to the allowable horizontal shear stress (F'_v), as expressed by:

$$\mathbf{f_v \le F'_v}$$

Determine the Allowable Horizontal Shear Stress (F'_v)

See Assumptions: 170 psi

Determine the Maximum Horizontal Shear Stress (f_v)

$$\begin{aligned} \mathbf{f_v} &= \mathbf{1.5\, V / A} \\ &= 1.5 \times (2.88 \times 1{,}000) / 63.25 \\ &= 68.30 \text{ lb/in}^2 \end{aligned}$$

where:
f_v = maximum horizontal shear stress parallel to grain
V = maximum vertical shear force = 2.88 k
A = area of cross section = 63.25 in^2

(for unit consistency, V is multiplied by 1,000 to convert k to lbs)

Shape	Maximum Horizontal Shear Stress (f_v)	Allowable Horizontal Shear Stress (F'_v)
6×12	68.30 psi	170 psi

Since the maximum horizontal shear stress is less than the allowable horizontal shear stress, the 6×12 passes the check for horizontal shear.

Step 3: Checking for Deflection (Beam 3)

For the 24 ft long 6×12 beam, assure that its maximum deflection is less than or equal to its allowable deflection for both:

- live loads (only)
- total loads (dead + live)

Live Load Deflection

- Live Load: maximum Deflection for the 6×12 beam

$$\boldsymbol{\delta_{LL} = 5\left(w_{LL}L \times L^3\right) / 384\,E'I}$$

where:
δ_{LL} = live load maximum deflection (in)
w_{LL} = uniform live load = 0.16 k/ft = 160 lb/ft
L = length = 24 ft
E′ = adjusted modulus of elasticity = 1,700,000 lb/in^2
I_{xx} = moment of inertia = 697.1 in^4

$$\delta_{LL} = \left[(5)\times(160\times24)\times(24\times12)^3\right] / (384\times1{,}700{,}000\times697.1) = 1.01 \text{ in}$$

- Live Load: Allowable Deflection for the 6×12 beam

$$\boldsymbol{\Delta_{LL} = L / 360}$$

$$\Delta_{LL} = (24\times12)/360 = 0.80 \text{ in}$$

where:
Δ_{LL} = live load allowable deflection (in)

Shape	Live Load Maximum Deflection (δ_{LL})	Live Load Allowable Deflection (Δ_{LL})
6×12	1.01 in	0.80 in

Since the live load maximum deflection is greater than the live load allowable deflection, the 6×12 does NOT pass the check for live load deflection. Return to Table 1B and select the next deeper shape 6×14 with an I_{xx} of 1,128 in^4.

(Note that if there was a depth restriction, the designer could also have selected, and checked for, the next wider shape 8×12.)

- Live Load: Maximum Deflection for the 6×14 beam

$$\delta_{LL} = \left[(5)\times(160\times 24)\times(24\times 12)^3\right]/(384\times 1{,}700{,}000\times 1{,}128) = 0.62\text{ in}$$

Shape	Live Load Maximum Deflection (δ_{LL})	Live Load Allowable Deflection (Δ_{LL})
6×14	0.62 in	0.80 in

Since the live load maximum deflection is less than the live load allowable deflection, the 6×14 passes the check for live load deflection.

Total Load Deflection

- Total Load: Maximum Deflection for the 6×14 beam

$$\boldsymbol{\delta_{TL} = 5\left(w_{TL}L\times L^3\right)/\,384\,E'I}$$

where:
δ_{TL} = total load maximum deflection (in)
w_{TL} = uniform total load = 0.24 k/ft = 240 lb/ft
L = length = 24 ft
E′ = adjusted modulus of elasticity = 1,700,000 lb/in²
I = moment of inertia = 1,128 in⁴

$$\delta_{TL} = [(5)\times(240\times 24)\times(24\times 12)^3]/(384\times 1{,}700{,}000\times 1{,}128) = 0.94\text{ in}$$

- Total Load: Allowable Deflection for the 6×14 beam

$$\boldsymbol{\Delta_{TL} = L/240}$$
$$\Delta_{TL} = (24\times 12)/240 = 1.20\text{ in}$$

where:
Δ_{TL} = total load allowable deflection (in)

Shape	Total Load Maximum Deflection (δ_{TL})	Total Load Allowable Deflection (Δ_{TL})
6×14	0.94 in	1.20 in

Since the total load actual deflection is less than the total load allowable deflection, the 6×14 passes the check for total load deflection, so we'll confirm it as our selection for Beam 3.

8.3 GIRDER B

For the design of Girder B, we'll follow steps similar to those in Beam 3.

Determine Loads (Girder B)

Girder B is supporting three sets of equally spaced beams (six beams total), each beam having the same reactions as Beam 3. The combined point load (P) from each of these sets of beams is 5.76 k (i.e., the end reaction from each beam; 2.88 k + 2.88 k) (Figure 8.7).

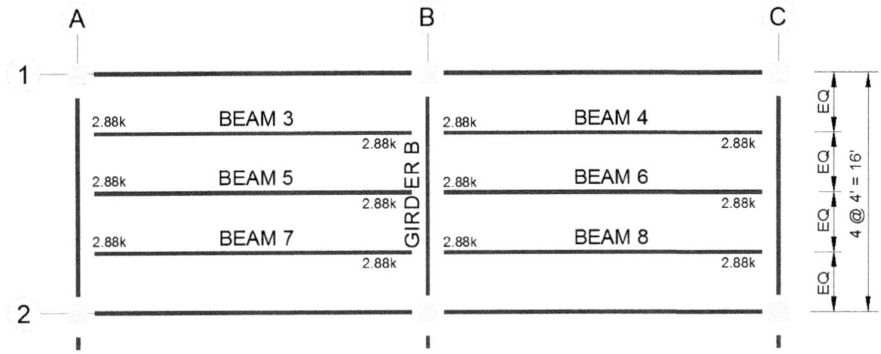

Figure 8.7 Loads on Girder B

Determine Reactions, Maximum Shear, Maximum Moment (Girder B)

Girder B is simply supported with three equally spaced point loads. From Appendix 4—Beam Diagrams and Formulae—the shear and moment diagrams with maximum values for Girder B are shown in Figure 8.8.

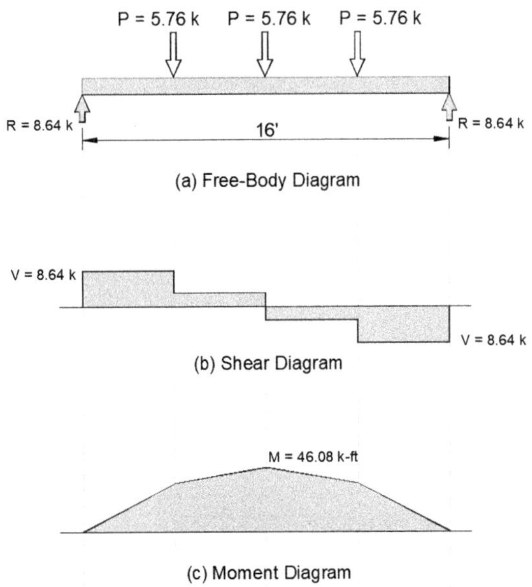

Figure 8.8 Girder B: Free-body, Shear & Moment Diagrams

$L\ (\text{length}) = 16\ \text{ft}$

$P\ (\text{point load}) = 2 \times 2.88\ \text{k} = 5.76\ \text{k}$

Reactions

$R = 3P/2$

$R = (3 \times 5.76)/2$

$\mathbf{R = 8.64\ k}$

Maximum Shear

$V = R$

$\mathbf{V = 8.64\ k}$

Maximum Moment

$M = 0.5PL$

$M = 0.5 \times 5.76 \times 16$

$\mathbf{M = 46.08\ k\text{-}ft}$

DESIGN OF GIRDER B

Step 1: Resisting Moment (Girder B)

Determine the Allowable Bending Stress (F'_b)

See Assumptions: 1,900 psi

Calculate the Required Section Modulus (S)

$$\begin{aligned} \mathbf{S} &= \mathbf{M} / \mathbf{F'_b} \\ &= (46.08 \times 2 \times 1{,}000) / 1{,}900 \\ &= 291.03 \text{ in}^3 \end{aligned}$$

where:
S = required section modulus (in^3)
M = maximum moment = 46.08 k-ft
F'_b = allowable bending stress = 1,900 lb/in^2

(for unit consistency, M is multiplied by 12 to convert ft to inches, and by 1,000 to convert k to lbs)

Select the Girder from Tables

From Table 1B in Appendix 2 (Section Properties of Standard Dressed [S4S] Sawn Lumber), select a shape with an available S_{xx} greater than that required.

Let's consider two viable options:

Nominal Beam Size (b × d)	S_{xx} (in^3)	Area (in^2)	Required Section Modulus (S)
8×16	300.3	116.3	291.03 in^3
6×20	330.9	118.3	

Again, in the interest of minimizing floor-to-floor heights, let's tentatively select the shallower 8×16 (with I_{xx} of 2327 in^4 and cross-sectional area A of 116.3 in^2), pending a shear and deflection check.

For the 8×16, let's compare the actual bending stress to the allowable bending stress.

$$\begin{aligned} \mathbf{f_b} &= \mathbf{M} / \mathbf{S} \\ &= (46.08 \times 12 \times 1{,}000) / 300.3 \\ &= 1841.4 \text{ psi} \end{aligned}$$

Actual Bending Stress (f_b)	Allowable Bending Stress (F'_b)
1,841.4 psi	1,900 psi

We see that the actual bending stress is less than the allowable.

Step 2: Checking for Horizontal Shear (Girder B)

Checking the 8×16 wood beam for horizontal shear involves assuring that the maximum horizontal shear stress (f_v) is less than or equal to the allowable horizontal shear stress (F'_v).

$$\mathbf{f_v \le F'_v}$$

Determine the Allowable Horizontal Shear Stress (F'_v)

See Assumptions: 170 psi

Determine the Maximum Horizontal Shear Stress (f_v)

$$\begin{aligned} f_v &= \mathbf{1.5\,V / A} \\ &= 1.5 \times (8.64 \times 1{,}000) / 116.3 \\ &= 111.5 \text{ psi} \end{aligned}$$

where:
f_v = maximum horizontal shear stress parallel to grain
V = maximum vertical shear force = 8.64 k
A = area of cross section = 116.3 in^2

(for unit consistency, V is multiplied by 1,000 to convert k to lbs)

Shape	Maximum Horizontal Shear Stress (f_v)	Allowable Horizontal Shear Stress (F'_v)
8×16	111.5 psi	170 psi

Since the maximum horizontal shear stress is less than the allowable horizontal shear stress, the 8×16 passes the check for horizontal shear.

Step 3: Checking for Deflection (Girder B)

For the 16 ft long 8×16 girder, assure that its maximum deflection is less than or equal to its allowable deflection for both:

- live loads (only)
- total loads (dead + live)

The point total (dead + live) load is 5.76 k. Since the ratio of live load to total load is 40 psf to 60 psf (i.e., 0.667):

$$P_{TL} = \text{point total load} = 5.76 \text{ k}$$
$$P_{LL} = \text{point live load} = 5.76 \times 0.667 = 3.84 \text{ k}$$

Live Load Deflection

- Live Load: Maximum Deflection for the 8×16 girder

$$\begin{aligned} \delta_{LL} &= \mathbf{0.05\,P_{LL}L^3 / E'I} \\ &= [0.05 \times 3{,}840 \times (16 \times 12)^3)] \\ &\quad / (1{,}700{,}000 \times 2{,}327) \\ &= 0.34 \text{ in} \end{aligned}$$

where:
δ_{LL} = live load maximum deflection (in)
P_{LL} = point live load = 3,840 lb
L = length = 16 ft
E′ = adjusted modulus of elasticity = 1,700,000 lb/in^2
I = moment of inertia = 2,327 in^4

- Live Load: Allowable Deflection for the 8×16 girder

$$\Delta_{LL} = L / 360$$
$$= (16 \times 12) / 360$$
$$= 0.53 \text{ in}$$

where:
Δ_{LL} = live load allowable deflection (in)

Section	Live Load Maximum Deflection (δ_{LL})	Live Load Allowable Deflection (Δ_{LL})
8×16	0.34 in	0.53 in

Since the live load actual deflection is less than the live load allowable deflection, the 8×16 passes the check for live load deflection.

Total Load Deflection

- Total Load: Maximum Deflection for the 8×16 girder

$$\delta_{TL} = 0.05\, P_{TL} L^3 / E'I$$

where:
δ_{LL} = total load maximum deflection (in)
P_{TL} = point live load = 5,760 lb
L = length = 16 ft
E′ = adjusted modulus of elasticity = 1,700,000 psi
I = moment of inertia = 2,327 in^4

$$\delta_{TL} = \left[0.05 \times 5{,}760 \times (16 \times 12)^3\right] / (1{,}700{,}000 \times 2{,}327) = 0.51 \text{ in}$$

- Total Load: Allowable Deflection for the 8×16 girder

$$\Delta_{TL} = L / 240$$
$$= (16 \times 12) / 240 = 0.80 \text{ in}$$

where:
Δ_{TL} = total load allowable deflection (in)

Section	Total Load Maximum Deflection (δ_{TL})	Total Load Allowable Deflection (Δ_{TL})
8×16	0.80 in	0.51 in

Since the total load maximum deflection is less than the total load allowable deflection, the 8×16 passes the check for total load deflection, so we'll confirm it as our selection for Girder B.

8.4 COLUMN B2

Determine Loads

First Floor Loads

Column B2 is supporting first floor Girders B and E and Beams 9 and 10. The combined load from the end reaction of each of these members is tabulated below and shown in Figure 8.9.

Member	First Floor Load on Column B2
Girder B	8.64 k
Girder E	8.64 k
Beam 9	2.88 k
Beam 10	2.88 k
Total	23.04 k

Roof Loads

In addition to the first floor loads, Column B2 is also supporting roof loads. Since the roof framing and loading is symmetrical about Column B2, we can calculate its roof load based on its load tributary area (Figure 8.10).

$$\text{Load tributary area for Column B2} = 16 \text{ ft} \times 24 \text{ ft}$$
$$= 384 \text{ sf}$$

D = uniform roof dead load = 20 psf

L_r = uniform roof live load = 30 psf

Applying the controlling Allowable Stress Design roof load combination:

$$D + L_r = 20 + 30 = 50 \text{ psf (see Chapter 4)}$$

$$\text{Roof load on Column B2} = 384 \text{ ft}^2 \times 50 \text{ lb/ft}^2$$
$$= 19{,}200 \text{ lb} = 19.20 \text{ k}$$

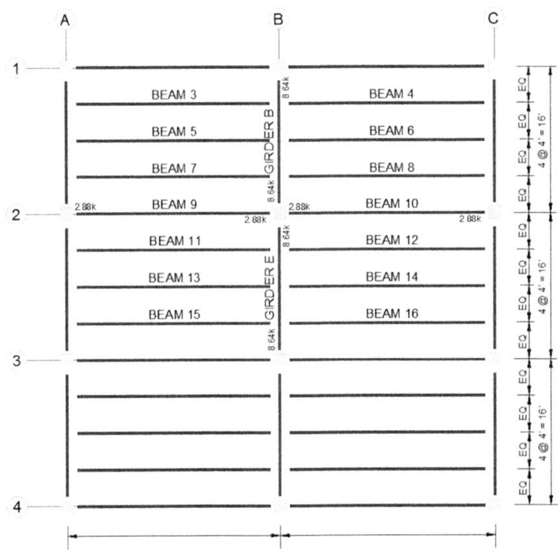

Figure 8.9 First Floor Loads on Column B2

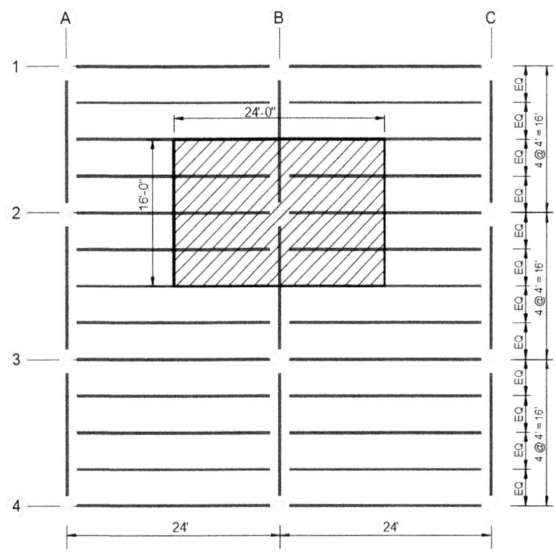

Figure 8.10 Roof Tributary Area for Column B2

Total Floor and Roof Loads on Column B2

First floor load	= 23.04 k
First floor load	= 23.04 k

Total Load on Column B2 = 42.24 k (required compressive strength)

Column Assumptions

Species, Grade, and Adjustment Factors

- Species: Douglas Fir-Larch
- Commercial grade: Dense Select Structural
- F_c = 1,350 psi
- E_{min}' = 620,000 psi
- Applicable adjustment factors (except C_p) = 1

Column Dimensions

Section size: 8×8 (nominal)

Dressed size: 7.5″ × 7.5″ (actual)

Section area = 56.25 in² (see NDS Table 1B)

Determine Effective Length and Slenderness Ratio

Effective Length

$$L_e = kl = 1.0 \times 12 = 12\text{ ft} \times 12 = 144\text{ in}$$

Slenderness Ratio

For rectangular solid wood columns, the NDS simplifies slenderness ratio as:

$$\textbf{slenderness ratio} = \mathbf{L_e / d} = 144/7.5 = 19.2$$

where:
L_e = effective length (in)
d = least (dressed) cross-section dimension = 7.5 in

Since the slenderness ratio does not exceed 50, the assumed section is acceptable.

Determine Allowable Compressive Stress (F'_c)

The allowable compressive stress is given by:

$$\mathbf{F'_c = F_c\left(C_D \times C_M \times C_t \times C_F \times C_i \times C_p\right)}$$

where:
F'_c = allowable compressive stress parallel to grain
F_c = reference compressive design value parallel to grain
$C_{(D,\,M,\,t,\,F,\,i)}$ = 1 (see Table 7.1)
C_p = column stability factor

Calculate the column stability adjustment factor (Cp)

$$C_p = \frac{1+\left(F_{cE}/F_c^*\right)}{2c} - \sqrt{\left[\frac{1+\left(F_{cE}/F_c^*\right)}{2c}\right]^2 - \frac{F_{cE}/F_c^*}{c}}$$

where:
C_p = column stability adjustment factor
F_{cE} = critical buckling design value
F^*_c = modified reference compressive design value parallel to grain, multiplied by all applicable adjustment factors (except C_p)
c = 0.8

Solve for (F_{cE})

$$F_{cE} = \frac{0.822\ E_{min}'}{\left(\frac{le}{d}\right)^2}$$

where:
E_{min}' = 620,000 lb/in^2
L_e/d = 19.2

$$F_{cE} = (0.822 \times 620{,}000)/(19.2)^2 = 1{,}382.5\ \text{lb/in}^2$$

Solve for (F^*_c)

$$F^*_c = F_c \times (C_D \times C_M \times C_t \times C_F \times C_i)$$
$$= 1{,}350 \times (1 \times 1 \times 1 \times 1 \times 1) = 1{,}350\ \text{lb/in}^2$$

where:
F_c = reference compressive design value parallel to grain (1,350 psi)
$C_{(D,\ M,\ t,\ F,\ i)}$ = applicable adjustment factors

Solve for (F_{cE}/F^*_c)

$$F_{cE}/F^*_c = 1{,}382.5/1{,}350 = 1.024$$

Solve for (C_p)

$$C_p = \frac{(1+1.024)}{2\times0.8} - \sqrt{\left[\frac{1+1.024}{2\times0.8}\right]^2 - \frac{1.024}{0.8}} = 0.686$$

The allowable compressive stress (F'_c) is determined to be:

$$F'_c = F_c(C_D \times C_M \times C_t \times C_F \times C_i \times C_p) = 1{,}350(1\times1\times1\times1\times1\times0.686) = 926.1\ \text{lb/in}^2$$

Determine Column Capacity

$$\mathbf{P} = \mathbf{F'_c} \times \mathbf{A}$$

$$= 926.1 \times 56.25$$
$$= 52{,}094 \text{ lb}$$
$$= 52.1 \text{ k}$$

where:
P = column capacity
F'_c = allowable compressive stress = 926.1 lb/in^2
A = cross-sectional area = 56.25 in^2

Verify Column Capacity ≥ Required Compressive Strength

Column Capacity	Required Compressive Strength
52.1 k	42.24 k

Since the column capacity is greater than the required compressive strength, the 8×8, Douglas Fir-Larch, No. 1 grade will be used for Column B2.

9 Design in Engineered Wood—Case Study

Our Case Study for design in engineered wood is a two-story one-way framed structure, 48 ft × 48 ft, with closely spaced joists spanning 24 ft, girders spanning 16 ft, and 12 ft floor-to-floor heights. We'll focus on the structural design of the first floor and select a typical joist, Girder B, and Column B2 to design using manufacturer-provided design criteria and tables. Manufacturers base their design criteria on Allowable Stress Design and guidelines in the National Design Specification (NDS) Package for Wood Construction. In addition to the loads from the first floor, the loads from the roof will be added onto Column B2. Figure 9.1 shows the first floor faming plan.

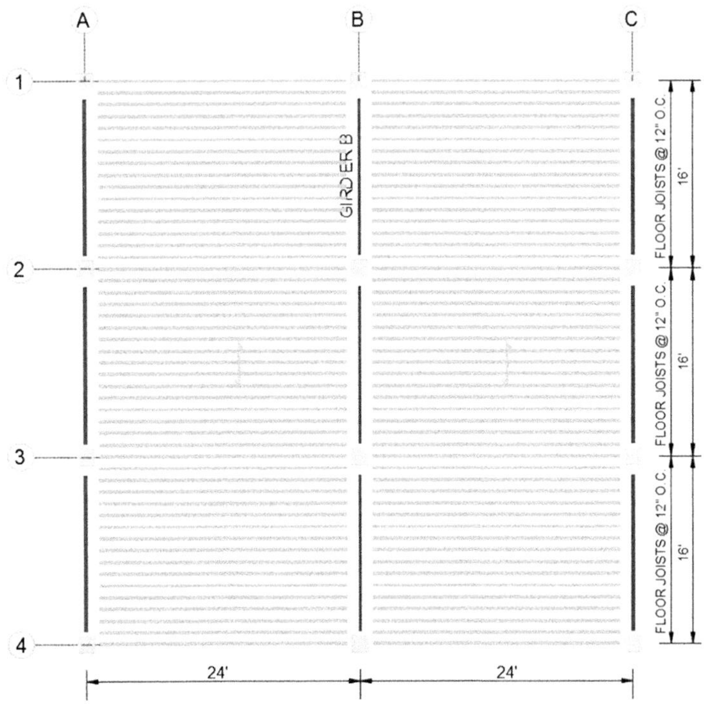

Figure 9.1 First Floor Framing Plan

9.1 ASSUMPTIONS

Floor Construction

Floor construction will be 3/4″ plywood subfloor, supported on wood joists 12″ on center as a one-way spanning system (Figure 9.2).

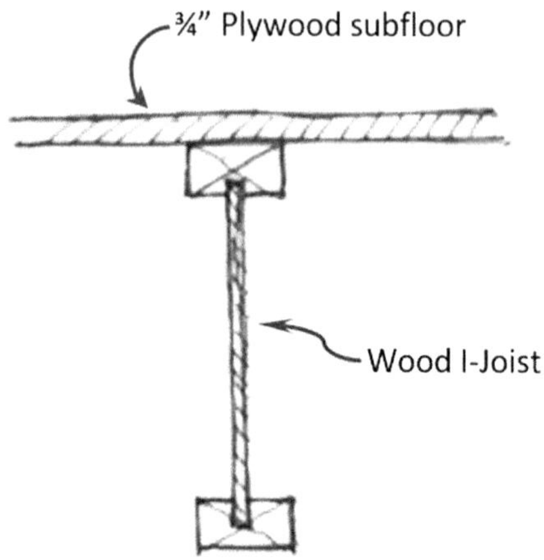

Figure 9.2 Floor Construction

Member Products and Shapes

Structural members will be manufactured I-joists and PSL girders and columns. The Weyerhaeuser Corporation is a well-established manufacturer of engineered wood products. TJI and Parallam are their trade names for I-joists and PSL sections respectively.

Floor and Roof Loads

We'll assume the following service loads:

Floor Loads		Roof Loads	
Dead Load		***Dead Load***	
*self-weight of beam, floor, and finishes	= 12 psf	*self-weight of roof framing and finishes	= 7 psf
mechanical (pipes, ducts, etc.)	= 8 psf	mechanical (pipes, ducts, etc.)	= 8 psf
Total	**= 20 psf**	**Total**	**= 15 psf**
Live Load		***Live Load***	
Total	**= 40 psf**	**Total**	**= 20 psf**

*In wood design, the self-weight of structural members is relatively small in relation to the other dead loads, and is typically included as a suitable square foot allowance in the dead load calculations.

Load Duration Adjustment Factor

Wood members can resist higher stresses over short periods and lower stresses over extended periods, accounted for by the load duration factor (C_D). The Weyerhaeuser design tables refer to 100% load durations, reflecting a load duration factor of 1.

Joists and Girders

Joists and girders will be considered to have:

- simple supports
- full lateral bracing

Columns

Columns will be considered to have:

- axial loads only, not subject to lateral loads
- floor-to-floor height (L) = 12 ft
- pinned top and bottom restraints; therefore, the k-factor = 1
  - effective length L_e = kL = 12 ft

Let's proceed to design a typical joist, Girder B, and Column B2. We'll use Weyerhaeuser's relevant product information and design tables that are included at the end of this chapter. These tables are based on Allowable Stress Design and take into account self-weight, allowable shear stresses, allowable deflection, and slenderness ratios of columns. We've numbered the tables for ease of reference. For consistency with terminology in the tables, we'll use the term "capacity" to reflect "available strength".

9.2 A TYPICAL JOIST

Determine Loads

For the given dead and live floor loads, calculate the service load imposed on a typical joist.

The load tributary area for a typical joist is a section of floor 1 ft × 24 ft (Figure 9.3). The uniformly distributed loads on a typical joist are:

$$\begin{aligned} w_{DL} &= \text{uniform dead load} = 20\ \text{lb/ft}^2 \times 1\ \text{ft} = 20\ \text{lb/ft} \\ w_{LL} &= \text{uniform live load} = 40\ \text{lb/ft}^2 \times 1\ \text{ft} = 40\ \text{lb/ft} \\ \hline w_{TL} &= \text{uniform total (service) load} = 60\ \text{lb/ft} \end{aligned}$$

Applying the governing Allowable Stress Design load combination (1.0 D + 1.0 L) to $w_{DL} + w_{LL}$ (see Chapter 4):

Total Design Load (w)

$$\begin{aligned} \mathbf{w} &= 1.0w_{DL} + 1.0w_{LL} \\ &= (1.0 \times 20) + (1.0 \times 0.40) \\ &= \mathbf{60\ lb/ft} \end{aligned}$$

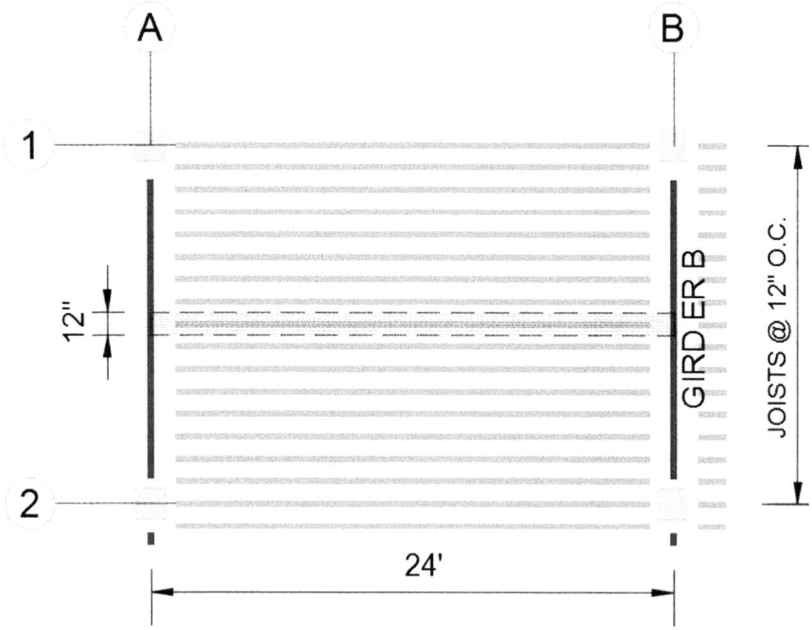

Figure 9.3 Load Tributary Area of a Typical Joist

Determine Reactions, Maximum Shear, Maximum Moment (Typical Joist)

A typical joist is simply supported with a uniformly distributed load along its length. Determine the reactions (R), maximum shear (V), and maximum moment (M). From Appendix 4—Beam Diagrams and Formulae—the shear and moment diagrams and reaction, shear, and moment values are as shown in Figure 9.4.

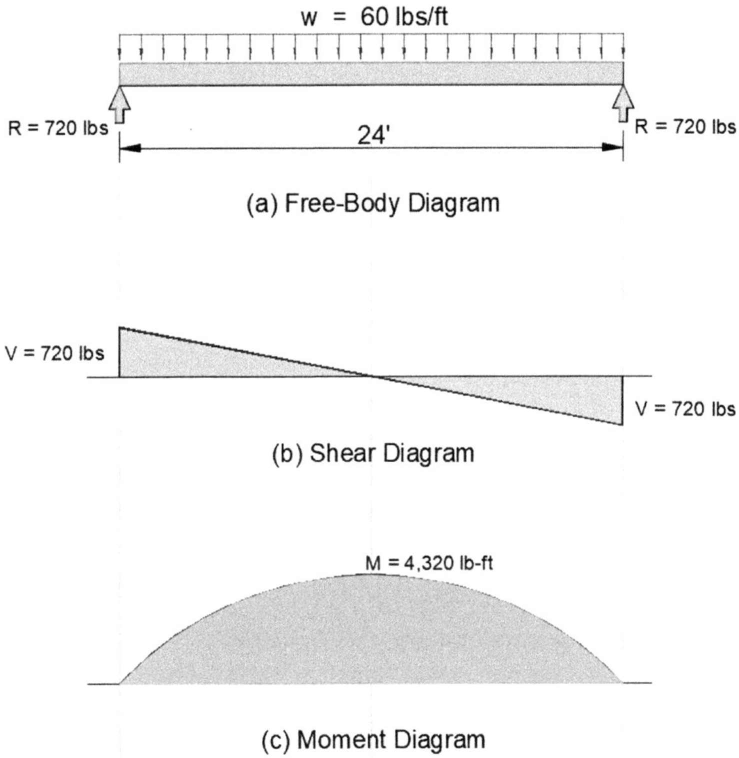

L (length) $= 24$ ft

$w = 60$ lb/ft

Reactions

$R = wL/2$

$R = (60 \times 24)/2$

$\mathbf{R = 720\ lb}$

Maximum (Actual) Shear

$V = R$

$\mathbf{V = 720\ lb}$

Maximum (Actual) Moment

$M = wL^2/8$

$M = (60 \times 24^2)/8$

$\mathbf{M = 4{,}320\ lb\text{-}ft}$

Figure 9.4 Typical Joist: Free-body, Shear & Moment Diagrams

Selection of a Typical Joist

For the design of a typical joist, we'll use Weyerhaeuser's TJI Joist Specifier's Guide. From Table 9.2 (Floor Span Tables and Material Weights), follow the recommendations for "How to Use These Tables".

1. Determine the appropriate live load deflection criteria.

The table provides two deflection criteria from which to choose—L/480 and L/360. L/360 is the maximum deflection allowed by code. L/480 is a more restrictive (i.e., allowing lesser deflection) option. We'll use L/360 as our deflection criteria.

2. *Identify the live and dead load condition.*

The table provides two live and dead load conditions from which to choose:

- 40 psf Live Load / 10 psf Dead Load
- 40 psf Live Load / 20 psf Dead Load

We'll use the "40 psf Live Load / 20 psf Dead Load" tabular column since these are our floor loads.

3. *Select the on-center spacing.*

16″ on-center (o.c.) is a common joist spacing. Closer spacing results in lighter joists, farther spacing results in heavier joists. We'll select 12″ o.c. spacing.

4. *Scan down the table until you meet or exceed the span of your application.*

The joist span is 24 ft. Scanning down the 12″ o.c. tabular column, we see that 25′-4″ exceeds our span.

5. *Select the TJI joist and depth.*

For the above criteria, the table provides the selection of a TJI 360 with a depth of 11⅞″, and a TJI 210 with a depth of 14″. In the interest of reducing floor-to-floor heights, we'll select the TJI 360.

From Table 9.1 (Design Properties), for the TJI 360 with a depth of 11⅞″ we see that:

- Maximum Resistive Moment (i.e., moment capacity) = 6,180 ft-lbs
- Maximum Vertical Shear (i.e., shear capacity) = 1,705 lbs

Member	Maximum Resistive Moment (moment capacity)	Actual Moment	Maximum Vertical Shear (shear capacity)	Actual Vertical Shear
TJI 360 11⅞″ depth	6,180 ft-lbs	4,320 ft-lbs	1,705 lbs	720 lbs

Since the moment and shear capacities are greater than the actual moment and vertical shear, we can feel confident of our joist selection.

9.3 GIRDER B

Determine Loads

Girder B is supporting joists 12″ o.c. on each side. The combined point load from each set these joists is 1,440 lb (i.e., 720 lb + 720 lb). In practical terms, this can be considered a uniformly distributed load (w_{TL}) of 1,440 lb/ft.

Determine Reactions, Maximum Shear, Maximum Moment (Girder B)

Girder B is simply supported with a uniformly distributed load along its length. Determine the reactions (R), maximum shear (V), and maximum moment (M). From Appendix 4—Beam Diagrams and Formulae—the shear and moment diagrams and reaction, shear, and moment values are as shown in Figure 9.5.

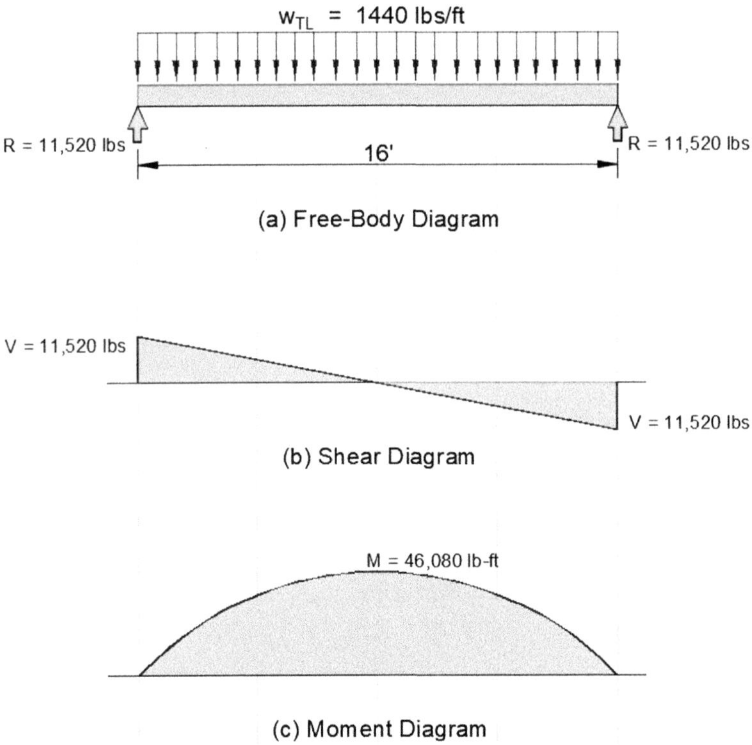

$L \text{ (length)} = 16 \text{ ft}$

$w_{TL} = 1{,}440 \text{ lb}$

Reactions

$R = wL/2$

$R = (1{,}440 \times 16)/2$

$\mathbf{R = 11{,}520 \text{ lb}}$

Maximum (Actual) Shear

$V = R$

$\mathbf{V = 11{,}520 \text{ lb}}$

Maximum (Actual) Moment

$M = w_{TL} L^2/8$

$M = (1{,}440 \times 16^2)/8$

$\mathbf{M = 46{,}080 \text{ lb-ft}}$

Figure 9.5 Girder B: Free-body, Shear & Moment Diagrams

Girder B Selection

For the design of Girder B, we'll use Weyerhaeuser's Beam, Header, and Columns Specifier's Guide. From Table 9.3 (Floor Load Tables—2.0E Parallam PSL), follow the recommendations for "How to Use This Table". This table provides the allowable uniform load on the specified members. Since our joist spacing is 12″ o.c., we can consider the load on Girder B to be a uniformly distributed load. Similar to the design selection of a typical joist, Table 9.3 is applicable for an L/360 deflection criteria.

1. *Calculate the total and live load on the beam in pounds per linear foot (plf).*

In other words, calculate the uniform total and live load on Girder B:

- the uniform total load (w_{TL}) = 1,440 plf
- since the ratio of live load (40 psf) to total load (60 psf) = 40/60 = 0.667, the uniform live load (w_{LL}) = 1,440 × 0.667 = 960 plf

2. *Select appropriate span.*

Girder B's span is 16 feet. Use the greater span of 16′-6″ in the table.

3. *Scan horizontally to find the proper width, and a depth with a capacity that exceeds actual total and live loads.*

For a 16′-6″ Span, in the Total Load row, scan horizontally to 1,514 plf. This is the lowest value that exceeds our uniform total load (w_{TL}) of 1,440 plf. We also see that the Live Load value is 1,074 plf, exceeding our uniform live load (w_{LL}) of 960 plf.

The selected member for Girder B is therefore 2.0E Parallam PSL, 5¼″ wide × 16″ deep.

Member	Allowable Uniform Total Load (from table)	Actual Uniform Total Load (w_{TL})	Allowable Uniform Live Load (from table)	Actual Uniform Live Load (w_{TL})
2.0E Parallam PSL 5¼″ wide × 16″	1,514 plf	1,440 plf	1,074 plf	960 plf

4. *Review bearing length requirements to ensure adequacy.*

From the table, the minimum end bearing requirement is 3.2″. We'll verify Girder B's bearing adequacy after the selection of a column.

From Table 9.4 (Design Properties), for the 2.0E Parallam PSL, 5¼″ wide × 16″ deep, we see that:

- Allowable Design Moment (i.e., moment capacity) = 52,430 ft-lbs
- Allowable Design Shear (i.e., shear capacity) = 16,240 lbs

Member	Maximum Resistive Moment (moment capacity)	Actual Moment	Maximum Vertical Shear (shear capacity)	Actual Vertical Shear
2.0E Parallam PSL 5¼″ wide × 16″	52,430 ft-lbs	46,080 ft-lbs	16,240 lbs	11,520 lbs

Since the moment and shear capacities are greater than the actual moment and vertical shear, we can feel confident of our girder selection.

Note that Table 9.3 is useful when a girder has a uniformly distributed load such as that resulting from 12″ o.c. joist spacing. Had we selected joist spacing further apart, we would calculate maximum moment and select a girder from Table 9.4.

9.4 COLUMN B2

Determine Loads

Since the joists are closely spaced, and since the joists and girders are symmetrical about Column B2, the simplest method to determine both the first floor and roof loads on Column B2 is from their load tributary areas (Figure 9.6).

First Floor Loads

Load tributary floor area for Column B2

$$= 24\text{ ft} \times 16\text{ ft} = 384\text{ ft}^2$$

Floor load on Column B2

$$= 384\text{ ft}^2 \times 60\text{ lb/ft}^2 = 23{,}040\text{ lb}$$

Roof Loads

The roof dead (service) load $(0) = 15$ psf

The roof live (service) load $(6) = 20$ psf

Load tributary roof area for Column B2

$$= 24' \times 16' = 384\text{ ft}^2$$

Applying the governing Allowable Stress Design load combination (1.0 D + 1.0 L) (see Chapter 4):

$$(1.0 \times 15) + (1.0 \times 20) = 35\text{ psf}$$

Roof load on Column B2

$$= 384 \times 35 = 13{,}440\text{ lb}$$

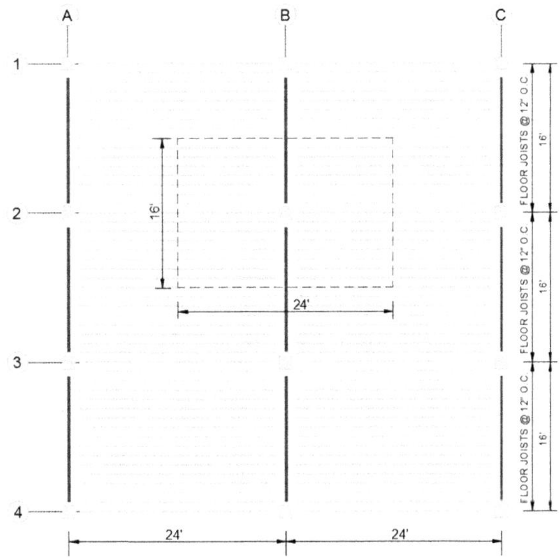

Figure 9.6 First Floor and Roof Load Tributary Area for Column B2

Total Floor and Roof Loads on Column B2

First floor load	= 23,040 lb
Roof load	= 13,440 lb
Total Load on Column B2	**= 36,480 lb** (actual axial load)

Column B2 Selection

From Table 9.5 (Columns—Allowable Axial Loads [lbs] for 1.8E Parallam PSL), select a suitable square section.

Design Criteria

- Column Bearing Type: On Column Base
- Effective Column Length = 12 ft

For the above criteria, we see that 7″ × 7″ is the smallest available square section having an allowable axial load greater than the actual axial load.

Member	Allowable Axial Load (axial capacity)	Actual Axial Load
1.8E Parallam PSL 7″ × 7″	40,000 lbs	36,480 lbs

We'll therefore select a 1.8E Parallam PSL, 7″ × 7″ for Column B2.

Review of Girder B's Bearing Length Requirements

Our design will assume girders sitting atop columns. With two girders sitting atop Column B2 (7″ × 7″), the length of Girder B's end bearing will be 3.5″ (1/2 of 7″).

Actual End Bearing	Minimum End Bearing
3.5″	3.2″

Since the actual end bearing is greater than the minimum end bearing requirement, Girder B has adequate bearing.

Table 9.1 Weyerhaeuser TJI Joist Specifier's Guide

DESIGN PROPERTIES

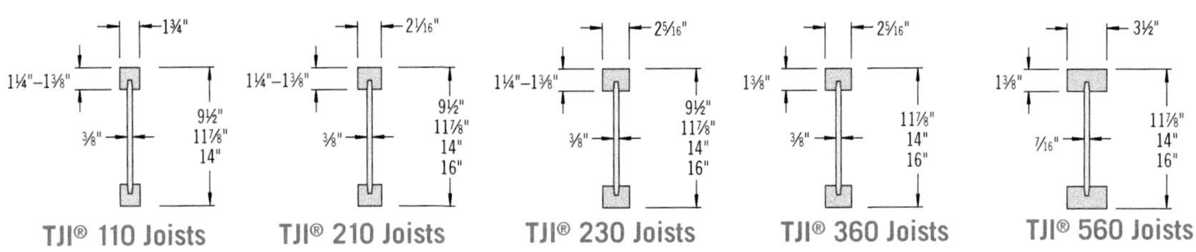

Design Properties (100% Load Duration)

Depth	TJI®	Basic Properties				Reaction Properties					
		Joist Weight (lbs/ft)	Maximum Resistive Moment(1) (ft-lbs)	Joist Only EI x 10^6 (in.^{2}-lbs)	Maximum Vertical Shear (lbs)	1¾" End Reaction (lbs)	3½" End Reaction (lbs)	3½" Intermediate Reaction (lbs)		5¼" Intermediate Reaction (lbs)	
								No Web Stiffeners	With Web Stiffeners(2)	No Web Stiffeners	With Web Stiffeners(2)
9½"	110	2.3	2,500	157	1,220	910	1,220	1,935	N.A.	2,350	N.A.
	210	2.6	3,000	186	1,330	1,005	1,330	2,145	N.A.	2,565	N.A.
	230	2.7	3,330	206	1,330	1,060	1,330	2,410	N.A.	2,790	N.A.
11⅞"	110	2.5	3,160	267	1,560	910	1,375	1,935	2,295	2,350	2,705
	210	2.8	3,795	315	1,655	1,005	1,460	2,145	2,505	2,565	2,925
	230	3.0	4,215	347	1,655	1,060	1,485	2,410	2,765	2,790	3,150
	360	3.0	6,180	419	1,705	1,080	1,505	2,460	2,815	3,000	3,360
	560	4.0	9,500	636	2,050	1,265	1,725	3,000	3,475	3,455	3,930
14"	110	2.8	3,740	392	1,860	910	1,375	1,935	2,295	2,350	2,705
	210	3.1	4,490	462	1,945	1,005	1,460	2,145	2,505	2,565	2,925
	230	3.3	4,990	509	1,945	1,060	1,485	2,410	2,765	2,790	3,150
	360	3.3	7,335	612	1,955	1,080	1,505	2,460	2,815	3,000	3,360
	560	4.2	11,275	926	2,390	1,265	1,725	3,000	3,475	3,455	3,930
16"	210	3.3	5,140	629	2,190	1,005	1,460	2,145	2,505	2,565	2,925
	230	3.5	5,710	691	2,190	1,060	1,485	2,410	2,765	2,790	3,150
	360	3.5	8,405	830	2,190	1,080	1,505	2,460	2,815	3,000	3,360
	560	4.5	12,925	1,252	2,710	1,265	1,725	3,000	3,475	3,455	3,930

(1) **Caution: Do not** increase joist moment design properties by a repetitive member use factor.

(2) See detail W on page 6 for web stiffener requirements and nailing information.

General Notes

- Design reaction includes all loads on the joist. Design shear is computed at the inside face of supports and includes all loads on the span(s). Allowable shear may sometimes be increased at interior supports in accordance with ICC ES ESR-1153, and these increases are reflected in span tables.
- The following formulas approximate the uniform load deflection of Δ (inches):

For TJI® 110, 210, 230, and 360 Joists

$$\Delta = \frac{22.5\, wL^4}{EI} + \frac{2.67\, wL^2}{d \times 10^5}$$

For TJI® 560 Joists

$$\Delta = \frac{22.5\, wL^4}{EI} + \frac{2.29\, wL^2}{d \times 10^5}$$

w = uniform load in pounds per linear foot
L = span in feet
d = out-to-out depth of the joist in inches
EI = value from table above

PRODUCT STORAGE

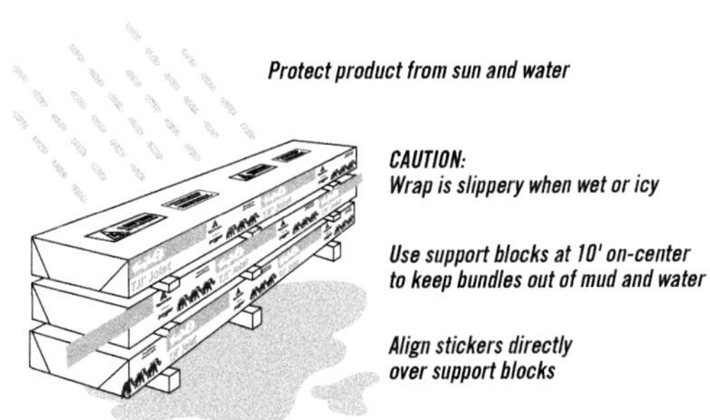

TJI® joists are intended for dry-use applications

Table 9.2 Weyerhaeuser TJI Joist Specifier's Guide

FLOOR SPAN TABLES AND MATERIAL WEIGHTS

L/480 Live Load Deflection

Depth	TJI®	40 PSF Live Load / 10 PSF Dead Load				40 PSF Live Load / 20 PSF Dead Load			
		12" o.c.	16" o.c.	19.2" o.c.	24" o.c.	12" o.c.	16" o.c.	19.2" o.c.	24" o.c.
9½"	110	16'-11"	15'-6"	14'-7"	13'-7"	16'-11"	15'-6"	14'-3"	12'-9"
	210	17'-9"	16'-3"	15'-4"	14'-3"	17'-9"	16'-3"	15'-4"	14'-0"
	230	18'-3"	16'-8"	15'-9"	14'-8"	18'-3"	16'-8"	15'-9"	14'-8"
11⅞"	110	20'-2"	18'-5"	17'-4"	15'-9"(1)	20'-2"	17'-8"	16'-1"(1)	14'-4"(1)
	210	21'-1"	19'-3"	18'-2"	16'-11"	21'-1"	19'-3"	17'-8"	15'-9"(1)
	230	21'-8"	19'-10"	18'-8"	17'-5"	21'-8"	19'-10"	18'-7"	16'-7"(1)
	360	22'-11"	20'-11"	19'-8"	18'-4"	22'-11"	20'-11"	19'-8"	17'-10"(1)
	560	26'-1"	23'-8"	22'-4"	20'-9"	26'-1"	23'-8"	22'-4"	20'-9"(1)
14"	110	22'-10"	20'-11"	19'-2"	17'-2"(1)	22'-2"	19'-2"	17'-6"(1)	15'-0"(1)
	210	23'-11"	21'-10"	20'-8"	18'-10"(1)	23'-11"	21'-1"	19'-2"(1)	16'-7"(1)
	230	24'-8"	22'-6"	21'-2"	19'-9"(1)	24'-8"	22'-2"	20'-3"(1)	17'-6"(1)
	360	26'-0"	23'-8"	22'-4"	20'-9"(1)	26'-0"	23'-8"	22'-4"(1)	17'-10"(1)
	560	29'-6"	26'-10"	25'-4"	23'-6"	***29'-6"***	***26'-10"***	25'-4"(1)	20'-11"(1)
16"	210	26'-6"	24'-3"	22'-6"(1)	19'-11"(1)	26'-0"	22'-6"(1)	20'-7"(1)	16'-7"(1)
	230	27'-3"	24'-10"	23'-6"	21'-1"(1)	***27'-3"***	23'-9"	21'-8"(1)	17'-6"(1)
	360	28'-9"	26'-3"	24'-8"(1)	21'-5"(1)	***28'-9"***	26'-3"(1)	22'-4"(1)	17'-10"(1)
	560	32'-8"	29'-8"	28'-0"	25'-2"(1)	***32'-8"***	***29'-8"***	26'-3"(1)	20'-11"(1)

L/360 Live Load Deflection (Minimum Criteria per Code)

Depth	TJI®	40 PSF Live Load / 10 PSF Dead Load				40 PSF Live Load / 20 PSF Dead Load			
		12" o.c.	16" o.c.	19.2" o.c.	24" o.c.	12" o.c.	16" o.c.	19.2" o.c.	24" o.c.
9½"	110	18'-9"	17'-2"	15'-8"	14'-0"	18'-1"	15'-8"	14'-3"	12'-9"
	210	19'-8"	18'-0"	17'-0"	15'-4"	19'-8"	17'-2"	15'-8"	14'-0"
	230	20'-3"	18'-6"	17'-5"	16'-2"	***20'-3"***	18'-1"	16'-6"	14'-9"
11⅞"	110	22'-3"	19'-4"	17'-8"	15'-9"(1)	20'-5"	17'-8"	16'-1"(1)	14'-4"(1)
	210	23'-4"	21'-2"	19'-4"	17'-3"(1)	***22'-4"***	19'-4"	17'-8"	15'-9"(1)
	230	24'-0"	21'-11"	20'-5"	18'-3"	***23'-7"***	20'-5"	18'-7"	16'-7"(1)
	360	25'-4"	23'-2"	21'-10"	20'-4"(1)	***25'-4"***	***23'-2"***	***21'-10"(1)***	17'-10"(1)
	560	28'-10"	26'-3"	24'-9"	23'-0"	***28'-10"***	***26'-3"***	***24'-9"***	20'-11"(1)
14"	110	24'-4"	21'-0"	19'-2"	17'-2"(1)	22'-2"	19'-2"	17'-6"(1)	15'-0"(1)
	210	26'-6"	23'-1"	21'-1"	18'-10"(1)	24'-4"	21'-1"	19'-2"(1)	16'-7"(1)
	230	27'-3"	24'-4"	22'-2"	19'-10"(1)	***25'-8"***	22'-2"	20'-3"(1)	17'-6"(1)
	360	28'-9"	26'-3"	24'-9"(1)	21'-5"(1)	***28'-9"***	***26'-3"(1)***	22'-4"(1)	17'-10"(1)
	560	32'-8"	29'-9"	28'-0"	25'-2"(1)	***32'-8"***	***29'-9"***	***26'-3"(1)***	20'-11"(1)
16"	210	28'-6"	24'-8"	22'-6"(1)	19'-11"(1)	26'-0"	22'-6"(1)	20'-7"(1)	16'-7"(1)
	230	30'-1"	26'-0"	23'-9"	21'-1"(1)	***27'-5"***	23'-9"	21'-8"(1)	17'-6"(1)
	360	31'-10"	29'-0"	26'-10"(1)	21'-5"(1)	***31'-10"***	***26'-10"(1)***	22'-4"(1)	17'-10"(1)
	560	36'-1"	32'-11"	31'-0"(1)	25'-2"(1)	***36'-1"***	***31'-6"(1)***	26'-3"(1)	20'-11"(1)

(1) Web stiffeners are required at intermediate supports of continuous-span joists when the intermediate bearing length is ***less*** than 5¼" and the span on either side of the intermediate bearing is greater than the following spans:

TJI®	40 PSF Live Load / 10 PSF Dead Load				40 PSF Live Load / 20 PSF Dead Load			
	12" o.c.	16" o.c.	19.2" o.c.	24" o.c.	12" o.c.	16" o.c.	19.2" o.c.	24" o.c.
110	Not Req.	Not Req.	Not Req.	15'-4"	Not Req.	Not Req.	16'-0"	12'-9"
210			21'-4"	17'-0"		21'-4"	17'-9"	14'-2"
230			Not Req.	19'-2"		Not Req.	19'-11"	15'-11"
360			24'-5"	19'-6"		24'-5"	20'-4"	16'-3"
560			29'-10"	23'-10"		29'-10"	24'-10"	19'-10"

- Long-term deflection under dead load, which includes the effect of creep, has not been considered. ***Bold italic*** spans reflect initial dead load deflection exceeding 0.33".

Material Weights

(Include TJI® weights in dead load calculations— see **Design Properties** table on page 3 for joist weights)

Floor Panels

Southern Pine

½" plywood .... 1.7 psf

⅝" plywood .... 2.0 psf

¾" plywood .... 2.5 psf

1⅛" plywood .... 3.8 psf

½" OSB .... 1.8 psf

⅝" OSB .... 2.2 psf

¾" OSB .... 2.7 psf

⅞" OSB .... 3.1 psf

1⅛" OSB .... 4.1 psf

Based on: Southern pine – 40 pcf for plywood, 44 pcf for OSB

Roofing

Asphalt shingles .... 2.5 psf

Wood shingles .... 2.0 psf

Clay tile .... 9.0 to 14.0 psf

Slate (⅜" thick) .... 15.0 psf

Roll or Batt Insulation (1" thick):

Rock wool .... 0.2 psf

Glass wool .... 0.1 psf

Floor Finishes

Hardwood (nominal 1") .... 4.0 psf

Sheet vinyl .... 0.5 psf

Carpet and pad .... 1.0 psf

¾" ceramic or quarry tile .... 10.0 psf

Concrete:

Regular (1") .... 12.0 psf

Lightweight (1") .... 8.0 to 10.0 psf

Gypsum concrete (¾") .... 6.5 psf

Ceilings

Acoustical fiber tile .... 1.0 psf

½" gypsum board .... 2.2 psf

⅝" gypsum board .... 2.8 psf

Plaster (1" thick) .... 8.0 psf

How to Use These Tables

1. Determine the appropriate live load deflection criteria.
2. Identify the live and dead load condition.
3. Select on-center spacing.
4. Scan down the column until you meet or exceed the span of your application.
5. Select TJI® joist and depth.

Live load deflection is not the only factor that affects how a floor will perform. To more accurately predict floor performance, use our TJ-Pro™ Ratings.

General Notes

- Tables are based on:
  - Uniform loads.
  - More restrictive of simple or continuous span.
  - Clear distance between supports
  - Minimum bearing length of 1¾" end (no web stiffeners) and 3½" intermediate.
- Assumed composite action with a single layer of 24" on-center span-rated, glue-nailed floor panels for deflection only. **Spans shall be reduced 6" when floor panels are nailed only.**
- Spans generated from Weyerhaeuser software may exceed the spans shown in these tables because software reflects actual design conditions.
- For multi-family applications and other loading conditions not shown, refer to Weyerhaeuser software or to the load table on page 5.

Trus Joist® TJI® **Joist Specifier's Guide** TJ-4000 | May 2013

4

Table 9.3 Weyerhaeuser Beam, Header, and Column Specifier's Guide

FLOOR LOAD TABLES

How to Use This Table

1. Calculate total and live load (neglect beam weight) on the beam or header in pounds per linear foot (plf).
2. Select appropriate **Span** (center-to-center of bearing).
3. Scan horizontally to find the proper width, and a depth with a capacity that exceeds actual total and live loads.
4. Review bearing length requirements to ensure adequacy.

*Also see **General Notes** on page 21.*

2.0E Parallam® PSL: Floor—100% (PLF)

Span	Condition	3½" Width							5¼" Width						
		9¼"	9½"	11¼"	11⅞"	14"	16"	18"	9¼"	9½"	11¼"	11⅞"	14"	16"	18"
8'	Total Load	1,469	1,517	1,861	1,990	2,456	2,933	2,933	2,204	2,275	2,792	2,985	3,683	4,400	4,400
	Live Load L/360	1,169	1,257	*	*	*	*	*	1,753	1,886	*	*	*	*	*
	Min. End/Int. Bearing (in.)	2.3/5.6	2.3/5.8	2.9/7.1	3.1/7.6	3.8/9.4	4.5/11.3	4.5/11.3	2.3/5.6	2.3/5.8	2.9/7.1	3.1/7.6	3.8/9.4	4.5/11.3	4.5/11.3
9'-6"	Total Load	1,076	1,147	1,510	1,611	1,970	2,333	2,467	1,614	1,720	2,265	2,416	2,955	3,500	3,700
	Live Load L/360	724	780	1,248	1,446	*	*	*	1,086	1,171	1,872	2,170	*	*	*
	Min. End/Int. Bearing (in.)	2.0/4.9	2.1/5.2	2.8/6.9	2.9/7.3	3.6/9.0	4.3/10.6	4.5/11.3	2.0/4.9	2.1/5.2	2.8/6.9	2.9/7.3	3.6/9.0	4.3/10.6	4.5/11.3
10'	Total Load	930	1,003	1,420	1,514	1,848	2,184	2,342	1,395	1,505	2,130	2,271	2,772	3,276	3,514
	Live Load L/360	626	675	1,084	1,257	*	*	*	940	1,013	1,626	1,886	*	*	*
	Min. End/Int. Bearing (in.)	1.8/4.5	1.9/4.8	2.7/6.8	2.9/7.3	3.5/8.9	4.2/10.5	4.5/11.3	1.8/4.5	1.9/4.8	2.7/6.8	2.9/7.3	3.5/8.9	4.2/10.5	4.5/11.3
12'	Total Load	548	592	964	1,092	1,480	1,738	1,949	822	888	1,446	1,639	2,220	2,607	2,923
	Live Load L/360	372	401	651	758	1,198	1,721	*	558	602	976	1,137	1,797	2,582	*
	Min. End/Int. Bearing (in.)	1.5/3.5	1.5/3.5	2.2/5.6	2.5/6.3	3.4/8.5	4.0/10.0	4.5/11.3	1.5/3.5	1.5/3.5	2.2/5.6	2.5/6.3	3.4/8.5	4.0/10.0	4.5/11.3
14'	Total Load	347	375	616	721	1,093	1,409	1,660	520	563	925	1,082	1,639	2,113	2,490
	Live Load L/360	238	257	419	489	780	1,132	1,561	357	386	629	734	1,171	1,698	2,342
	Min. End/Int. Bearing (in.)	1.5/3.5	1.5/3.5	1.7/4.2	2.0/4.9	3.0/7.4	3.8/9.5	4.5/11.3	1.5/3.5	1.5/3.5	1.7/4.2	2.0/4.9	3.0/7.4	3.8/9.5	4.5/11.3
16'-6"	Total Load	210	228	379	444	720	1,009	1,263	316	342	568	667	1,080	1,514	1,895
	Live Load L/360	147	159	260	305	490	716	995	220	238	391	457	735	1,074	1,493
	Min. End/Int. Bearing (in.)	1.5/3.5	1.5/3.5	1.5/3.5	1.5/3.6	2.3/5.8	3.2/8.1	4.0/10.1	1.5/3.5	1.5/3.5	1.5/3.5	1.5/3.6	2.3/5.8	3.2/8.1	4.0/10.1
18'-6"	Total Load	147	160	268	315	514	759	1,000	221	240	402	473	771	1,138	1,501
	Live Load L/360	105	113	186	218	352	517	722	157	170	280	328	529	776	1,084
	Min. End/Int. Bearing (in.)	1.5/3.5	1.5/3.5	1.5/3.5	1.5/3.5	1.9/4.7	2.7/6.8	3.6/9.0	1.5/3.5	1.5/3.5	1.5/3.5	1.5/3.5	1.9/4.7	2.7/6.8	3.6/9.0
20'	Total Load	115	125	210	248	407	603	850	172	187	316	372	610	905	1,275
	Live Load L/360	83	90	148	174	281	414	579	125	135	223	261	422	621	869
	Min. End/Int. Bearing (in.)	1.5/3.5	1.5/3.5	1.5/3.5	1.5/3.5	1.6/4.0	2.4/5.9	3.3/8.3	1.5/3.5	1.5/3.5	1.5/3.5	1.5/3.5	1.6/4.0	2.4/5.9	3.3/8.3
22'	Total Load	84	91	156	184	304	454	642	126	137	234	277	457	681	964
	Live Load L/360	63	68	112	131	213	314	441	94	102	168	197	320	472	662
	Min. End/Int. Bearing (in.)	1.5/3.5	1.5/3.5	1.5/3.5	1.5/3.5	1.5/3.5	2.0/4.9	2.8/6.9	1.5/3.5	1.5/3.5	1.5/3.5	1.5/3.5	1.5/3.5	2.0/4.9	2.8/6.9
24'	Total Load	62	68	118	140	232	349	496	94	103	177	210	349	523	744
	Live Load L/360	48	52	86	102	165	244	343	73	79	130	153	248	366	515
	Min. End/Int. Bearing (in.)	1.5/3.5	1.5/3.5	1.5/3.5	1.5/3.5	1.5/3.5	1.7/4.2	2.4/5.9	1.5/3.5	1.5/3.5	1.5/3.5	1.5/3.5	1.5/3.5	1.7/4.2	2.4/5.9
26'	Total Load		51	90	107	180	272	389	71	77	135	161	271	409	584
	Live Load L/360		41	68	80	130	193	272	57	62	102	120	196	290	409
	Min. End/Int. Bearing (in.)		1.5/3.5	1.5/3.5	1.5/3.5	1.5/3.5	1.5/3.6	2.0/5.1	1.5/3.5	1.5/3.5	1.5/3.5	1.5/3.5	1.5/3.5	1.5/3.6	2.0/5.1
28'	Total Load			70	84	142	216	310	54	59	105	126	213	324	465
	Live Load L/360			55	64	105	155	219	46	50	82	97	157	233	329
	Min. End/Int. Bearing (in.)			1.5/3.5	1.5/3.5	1.5/3.5	1.5/3.5	1.8/4.4	1.5/3.5	1.5/3.5	1.5/3.5	1.5/3.5	1.5/3.5	1.5/3.5	1.8/4.4
30'	Total Load			55	66	113	173	249			82	99	170	260	374
	Live Load L/360			44	52	85	127	179			67	79	128	190	269
	Min. End/Int. Bearing (in.)			1.5/3.5	1.5/3.5	1.5/3.5	1.5/3.5	1.5/3.9			1.5/3.5	1.5/3.5	1.5/3.5	1.5/3.5	1.5/3.9
32'	Total Load				52	91	140	203			64	78	136	210	305
	Live Load L/360				43	70	105	148			55	65	106	157	223
	Min. End/Int. Bearing (in.)				1.5/3.5	1.5/3.5	1.5/3.5	1.5/3.5			1.5/3.5	1.5/3.5	1.5/3.5	1.5/3.5	1.5/3.5

* Indicates **Total Load** value controls.

Table 9.4 Weyerhaeuser Beam, Header, and Column Specifier's Guide

DESIGN PROPERTIES

Allowable Design Properties(1) (100% Load Duration)

Grade	Width	Design Property	Depth												
			4⅜"	5½"	5½" Plank Orientation	7¼"	8⅝"	9¼"	9½"	11¼"	11⅞"	14"	16"	18"	20"
TimberStrand® LSL															
1.3E	3½"	Moment (ft-lbs)	1,735	2,685	1,780	4,550	6,335	7,242		10,521					
		Shear (lbs)	4,340	5,455	1,925	7,190	8,555	9,175		11,155					
		Moment of Inertia (in.⁴)	24	49	20	111	187	231		415					
		Weight (plf)	4.5	5.6	5.6	7.4	8.8	9.4		11.5					
1.55E	1¾"	Moment (ft-lbs)						4,950	5,210	7,195	7,975	10,920	14,090		
		Shear (lbs)						3,345	3,435	4,070	4,295	5,065	5,785		
		Moment of Inertia (in.⁴)						115	125	208	244	400	597		
		Weight (plf)						5.1	5.2	6.2	6.5	7.7	8.8		
	3½"	Moment (ft-lbs)						9,905	10,420	14,390	15,955	21,840	28,180		
		Shear (lbs)						6,690	6,870	8,140	8,590	10,125	11,575		
		Moment of Inertia (in.⁴)						231	250	415	488	800	1,195		
		Weight (plf)						10.1	10.4	12.3	13	15.3	17.5		
Microllam® LVL															
1.9E	1¾"	Moment (ft-lbs)		2,125		3,555		5,600	5,885	8,070	8,925	12,130	15,555	19,375	23,580
		Shear (lbs)		1,830		2,410		3,075	3,160	3,740	3,950	4,655	5,320	5,985	6,650
		Moment of Inertia (in.⁴)		24		56		115	125	208	244	400	597	851	1,167
		Weight (plf)		2.8		3.7		4.7	4.8	5.7	6.1	7.1	8.2	9.2	10.2
Parallam® PSL															
2.0E	3½"	Moment (ft-lbs)						12,415	13,055	17,970	19,900	27,160	34,955	43,665	
		Shear (lbs)						6,260	6,430	7,615	8,035	9,475	10,825	12,180	
		Moment of Inertia (in.⁴)						231	250	415	488	800	1,195	1,701	
		Weight (plf)						10.1	10.4	12.3	13.0	15.3	17.5	19.7	
	5¼"	Moment (ft-lbs)						18,625	19,585	26,955	29,855	40,740	52,430	65,495	
		Shear (lbs)						9,390	9,645	11,420	12,055	14,210	16,240	18,270	
		Moment of Inertia (in.⁴)						346	375	623	733	1,201	1,792	2,552	
		Weight (plf)						15.2	15.6	18.5	19.5	23.0	26.3	29.5	
	7"	Moment (ft-lbs)						24,830	26,115	35,940	39,805	54,325	69,905	87,325	
		Shear (lbs)						12,520	12,855	15,225	16,070	18,945	21,655	24,360	
		Moment of Inertia (in.⁴)						462	500	831	977	1,601	2,389	3,402	
		Weight (plf)						20.2	20.8	24.6	26.0	30.6	35.0	39.4	

(1) For product in beam orientation, unless otherwise noted.

Some sizes may not be available in your region.

PRODUCT STORAGE

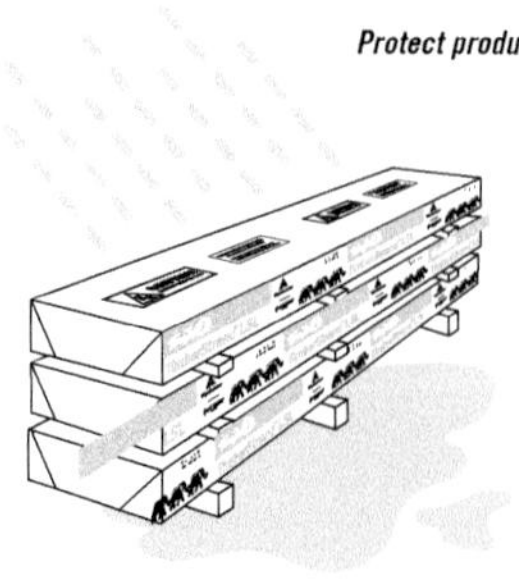

Table 9.5 Weyerhaeuser Beam, Header, and Column Specifier's Guide

COLUMNS

Allowable Axial Loads (lbs) for 1.3E TimberStrand® LSL

Column Bearing Type	Effective Column Length	Column Size														
		3½" x 3½"			3½" x 4⅜"			3½" x 5½"			3½" x 7¼"			3½" x 8⅝"		
		100%	115%	125%	100%	115%	125%	100%	115%	125%	100%	115%	125%	100%	115%	125%
On Column Base	3'	12,165	13,665	14,625	15,210	17,085	18,280	19,120	21,475	22,980	25,205	28,310	30,290	29,985	33,680	36,035
	4'	10,745	11,830	12,490	13,435	14,790	15,610	16,885	18,590	19,625	22,260	24,505	25,870	26,480	29,155	30,780
	5'	9,120	9,810	10,215	11,400	12,265	12,765	14,335	15,420	16,050	18,895	20,325	21,155	22,480	24,180	25,170
	6'	7,550	7,985	8,235	9,440	9,980	10,295	11,865	12,550	12,945	15,640	16,540	17,060	18,610	19,680	20,300
	7'	6,235	6,525	6,695	7,795	8,160	8,370	9,800	10,255	10,520	12,915	13,520	13,870	15,365	16,085	16,500
	8'	5,195	5,400	5,515	6,490	6,750	6,895	8,160	8,485	8,670	10,755	11,185	11,430	12,795	13,305	13,595
	9'	4,375	4,525	4,610	5,465	5,655	5,765	6,870	7,110	7,245	9,060	9,370	9,550	10,775	11,150	11,360
	10'	3,725	3,840	3,905	4,655	4,795	4,880	5,850	6,030	6,135	7,715	7,950	8,085	9,175	9,460	9,620
	12'	2,785	2,855	2,895	3,480	3,565	3,615	4,375	4,485	4,545	5,770	5,910	5,995	6,860	7,030	7,130
	14'	2,155	2,200	2,225	2,695	2,750	2,780	3,385	3,455	3,495	4,465	4,555	4,610	5,310	5,420	5,485
On Wood Plate(1)	3'–7'	5,765	5,765	5,765	7,065	7,065	7,065	8,740	8,740	8,740	10,785	10,785	10,785	12,830	12,830	12,830
	8'	5,195	5,400	5,515	6,490	6,750	6,895	8,160	8,485	8,670	10,755	10,785	10,785	12,795	12,830	12,830
	9'	4,375	4,525	4,610	5,465	5,655	5,765	6,870	7,110	7,245	9,060	9,370	9,550	10,775	11,150	11,360
	10'	3,725	3,840	3,905	4,655	4,795	4,880	5,850	6,030	6,135	7,715	7,950	8,085	9,175	9,460	9,620
	12'	2,785	2,855	2,895	3,480	3,565	3,615	4,375	4,485	4,545	5,770	5,910	5,995	6,860	7,030	7,130
	14'	2,155	2,200	2,225	2,695	2,750	2,780	3,385	3,455	3,495	4,465	4,555	4,610	5,310	5,420	5,485

(1) See connection details below.

Allowable Axial Loads (lbs) for 1.8E Parallam® PSL

Column Bearing Type	Effective Column Length	Column Size																	
		3½" x 3½"			3½" x 5¼"			3½" x 7"			5¼" x 5¼"			5¼" x 7"			7" x 7"		
		100%	115%	125%	100%	115%	125%	100%	115%	125%	100%	115%	125%	100%	115%	125%	100%	115%	125%
On Column Base	6'	10,595	11,200	11,545	15,890	16,800	17,320	21,190	22,395	23,095	33,295	36,675	38,735	40,000	40,000	40,000	40,000	40,000	40,000
	7'	8,735	9,140	9,370	13,105	13,710	14,060	17,475	18,280	18,745	30,010	32,545	34,030	40,000	40,000	40,000	40,000	40,000	40,000
	8'	7,265	7,550	7,715	10,900	11,325	11,570	14,535	15,100	15,425	26,650	28,490	29,555	35,530	37,985	39,410	40,000	40,000	40,000
	9'	6,115	6,320	6,440	9,170	9,480	9,660	12,225	12,640	12,880	23,475	24,835	25,620	31,300	33,115	34,165	40,000	40,000	40,000
	10'	5,200	5,355	5,445	7,800	8,035	8,170	10,400	10,715	10,895	20,660	21,695	22,290	27,545	28,925	29,725	40,000	40,000	40,000
	12'	3,885	3,980	4,030	5,825	5,965	6,050	7,765	7,955	8,065	16,160	16,805	17,175	21,545	22,405	22,900	40,000	40,000	40,000
	14'	3,000	3,065	3,100	4,500	4,595	4,645	6,005	6,125	6,195	12,890	13,315	13,560	17,185	17,755	18,080	34,155	35,785	36,720
	16'										10,480	10,775	10,950	13,970	14,370	14,595	28,485	29,640	30,300
	18'										8,670	8,885	9,010	11,560	11,850	12,010	24,020	24,860	25,345
	20'					Slenderness ratio exceeds 50					7,285	7,445	7,535	9,710	9,925	10,050	20,475	21,110	21,475
	22'																17,630	18,125	18,405
	24'																15,325	15,715	15,935

General Notes

- Tables are based on:
  - Solid, one-piece column members used in dry-service conditions.
  - Bracing in both directions at column ends.
  - NDS® 2005.
  - Simple columns with axial loads only. For side loads or other combined bending and axial loads, see the NDS® 2005.
- Wood plate bearing is based on compression perpendicular-to-grain stress of **425 psi adjusted per the NDS®** 2005, 3.10.4.
- Allowable loads have been adjusted to accommodate the worst case of the following eccentric conditions: ⅙ of column thickness (first dimension) or ⅙ of column width.
- Beams and columns must remain straight to within $\frac{5L^2}{4608}$ (in.) of true alignment. L is the unrestrained length of the member in feet.

For column allowable design stresses see page 5.

Top or Bottom Plate Connection

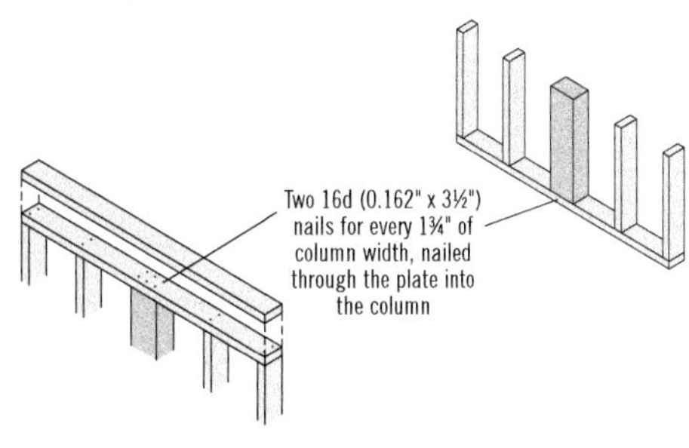

The column and connector values listed are for dry-service conditions ONLY. When wet-service conditions exist, contact your Weyerhaeuser representative for other product solutions.

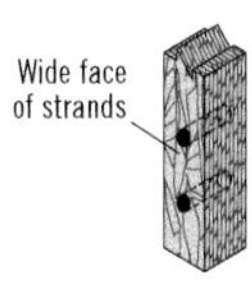

In order to use the manufacturer's published capacities when designing column caps, bases, or holdowns for uplift, the bolts or self-drilling screws must be installed perpendicular to the wide face of strands as shown at left.

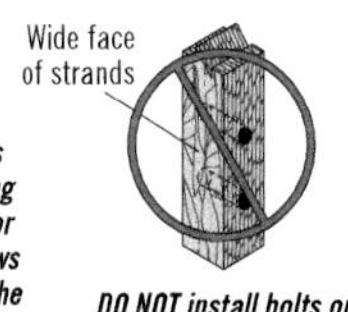

DO NOT install bolts or screws into the narrow face of strands

10 Understanding Reinforced Concrete

10.1 MATERIALS AND MANUFACTURE

Concrete is an artificial stone-like material made by mixing its three basic ingredients: cement, mineral aggregate, and water. In addition, various admixtures may be added to the concrete to modify its properties. Concrete is a universal construction material available worldwide. Concrete products can be produced on-site (variously referred to as *site-cast, cast-in-place*, or *in situ*), or produced off-site in sophisticated plants and shipped to the construction site. Either way, the manufacture of structural concrete is a carefully controlled process in which the proportions and quality of its ingredients are balanced for the optimum combination of desired physical properties, workability, and cost (Figure 10.1).

Figure 10.1 A Reinforced Concrete Frame Building

Basic Ingredients

Portland Cement—a manufactured product primarily consisting of lime, iron, silica, and alumina obtained by burning together several raw materials. Cement fills in the voids between the ingredients bonding them together into a rock-like mass.

Aggregate—consisting of clean, hard, sand, gravel, and crushed stone properly graded to minimize voids in the concrete mix. Aggregate constitutes approximately 75% of the volume of concrete and can vary according to the type of mineral material available in any particular locale. The typical maximum aggregate size is ¾″ to facilitate the flow of concrete around steel reinforcing.

Water—to activate the chemical reaction with cement resulting in the setting process of the cement mix with the ingredients. Water should be potable and free of organic material, clay, and salts. The compressive strength of concrete is inversely proportional to the ratio of water to cement in any particular concrete mix. In general, the less the water, the stronger the concrete (Figures 10.2 and 10.3).

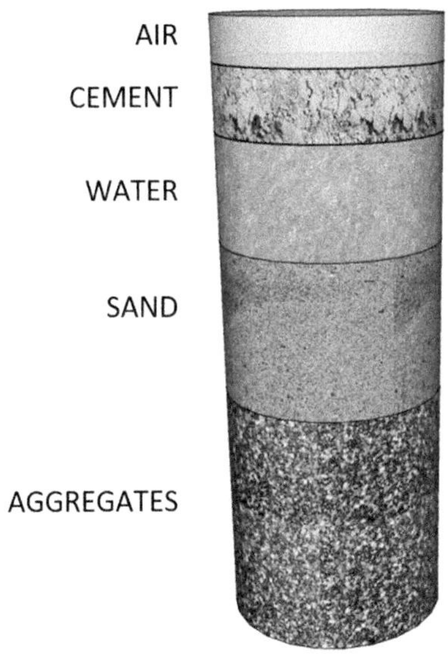

Figure 10.2 Approximate Ingredient Proportions of Concrete

Figure 10.3 Concrete Cylinder Core

Concrete's basic properties may be altered if desired, by the type of cement used and/or by adding admixtures to:

- increase workability and improve resistance to cracking
- reduce setting time and speed strength development
- slow the setting time and allow more time for placing and working the concrete
- reduce water content while maintaining workability
- inhibit rusting of steel reinforcing
- produce lightweight insulating concrete
- produce less heat when curing
- produce a desired color

Hydration and Curing

When water is added to a concrete mix, chemical reactions with the cement create a gel-like material that binds the aggregate. During the setting phase, concrete is commonly referred to as 'green' concrete. The process of cement reacting with water is called *hydration*.

The ratio of water to cement (termed water-cement ratio) is normally between 0.4 and 0.5 ensuring that adequate water is present for the complete hydration of the cement, and for the proper workability (Figure 10.4).

Figure 10.4 Pouring Concrete

Hydration is an *exothermic* process, meaning that it gives off heat. If not properly controlled, the heat can result in the water evaporating too quickly, creating a variety of problems, including excessive shrinkage cracks. For this reason, freshly poured concrete must be kept moist with temperatures maintained within specified limits while it is setting. This process, called *curing*, is vital for concrete to gain its proper strength and acquire other important properties (Figure 10.5).

Figure 10.5 Curing of Concrete

As concrete starts to set, it gains strength rapidly at first, but then the rate of strength-gain slows down. Concrete typically gains 50% of its strength within the first three days, and 75% of its strength within seven days. Although concrete theorctically continues to gain strength, its full design strength is assumed to be achieved at 28 days (Figure 10.6).

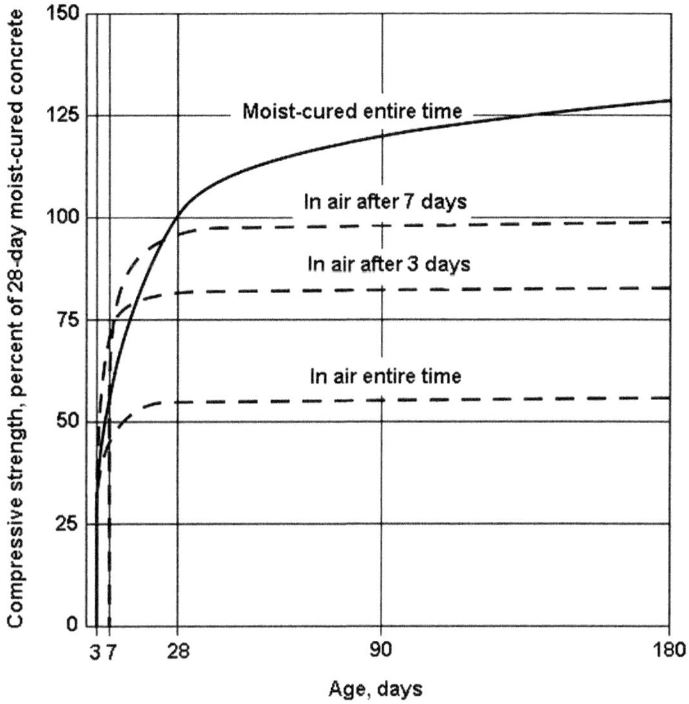

Figure 10.6 Concrete Curing Strength

Concrete shrinks as it sets. When set, temperature variations cause concrete to expand and contract creating stresses that tend to cause cracks. For slabs on grade these stresses are generally addressed by:

- reinforcement in the form of welded wire fabric (also known as wire mesh), steel reinforcing bars, or structural fibers
- control joints cut in concrete to induce cracking at pre-determined locations
- expansion joints placed at regular intervals to allow movement

(Figures 10.7–10.9)

Figure 10.7 Placing Steel Reinforcing in a Concrete Slab

Figure 10.8 Control Joints in a Concrete Slab

Figure 10.9 Expansion Joints in a Concrete Slab

10.2 STRUCTURAL CONSIDERATIONS

Concrete's structural usefulness is derived from its compressive strength.

Concrete Strength

Concrete is designed and specified as a function of its compressive strength, ranging from 2,500 to 5,000 psi for typical use. Concrete with higher compressive strengths is often used for larger structures, or for specific uses (such as the lower floors of a high-rise concrete building to keep column sizes small). A higher proportion of cement generally provides greater strength, but too much cement creates a condition whereby the excessive cement does not properly hydrate. In addition, more cement in a mix not only increases material cost, but also increases the heat of hydration, requiring stricter controls (and higher costs) for placement and curing.

Pozzolans

Pozzolans, termed supplementary cementitious materials (SCMs), commonly include fly ash, silica fume, and slag—all waste by-products of various industrial processes. As cement substitutes, pozzolans are introduced into a concrete mix to obtain higher strengths and reduce the quantity of cement.

- Fly ash is the by-product of burning coal in power generating plants (Figure 10.10).
- Silica-fume, also known as micro silica, is the by-product of the manufacture of silicon.
- Slag is the by-product of producing iron in blast furnaces.

Figure 10.10 Stockpiled Fly Ash

With the use of pozzolans, it is possible to obtain concrete strengths up to about 19,000 psi.

Steel Reinforcing

Without steel reinforcing, concrete is virtually useless in tension. In order to resist tensile stresses in concrete members, steel reinforcing bars (commonly called 'rebar') are added—resulting in *reinforced concrete* (Figure 10.11).

Figure 10.11 Steel Rebar

- Rebar cross-sectional diameters range from 3/8″ to 1″, in 1/8″ increments. Within this range, bar thickness is identified by a number that corresponds to its cross-sectional diameter. For example, a #3 bar has a 3/8″ diameter; a #7 bar has a 7/8″ diameter. Larger bars (#9, 10, 11, 14, and 18) are also available with increasing diameters.
- The most commonly used grade of steel for reinforcing bars is Grade 60 with a yield strength of 60 ksi. Other common grades are Grades 40 and 75, with yield strengths of 40 and 75 ksi respectively.
- Surface deformations (or ridges) on the rebar help it bond with the concrete (referred to as 'deformed' rebar).
- Epoxy-coated or galvanized rebars are used for increased resistance to corrosion.
- Stainless steel rebars, though more expensive, are highly resistant to corrosion and are often specified where exposure or corrosiveness is severe.

Modulus of Elasticity

The ACI provides the following formula to calculate modulus of elasticity.

$$\mathbf{E_c\,(psi) = 5{,}700\ \sqrt{f}_{c}}$$

where:
E_c = modulus of elasticity of concrete
f'_c = 28-day strength of concrete in psi

10.3 DESIGN CONSIDERATIONS FOR BEAMS

Bending in a Reinforced Concrete Beam

In a simple beam under gravity loads, bending will create compressive stresses above the neutral axis, tensile stresses below the neutral axis, and shear stresses along the length of the beam. (Figure 10.12).

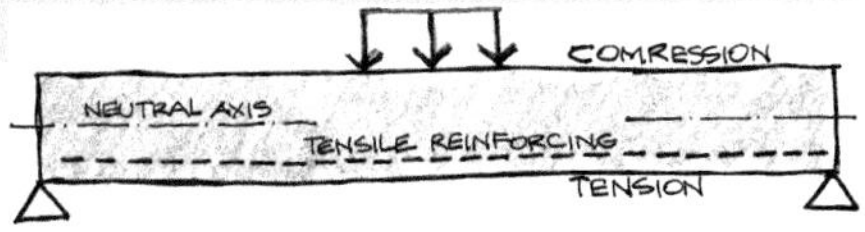

Figure 10.12 Tensile Reinforcing in a Simply Supported Rrinforced Concrete Beam

Compressive stresses are resisted by the concrete above the neutral axis while tensile stresses are resisted by steel tensile reinforcing below the neutral axis (Figures 10.13a and b). Shear stresses are resisted by the entire section of concrete, and the shear strength of the section is supplemented by steel shear reinforcing, commonly in the form of stirrups.

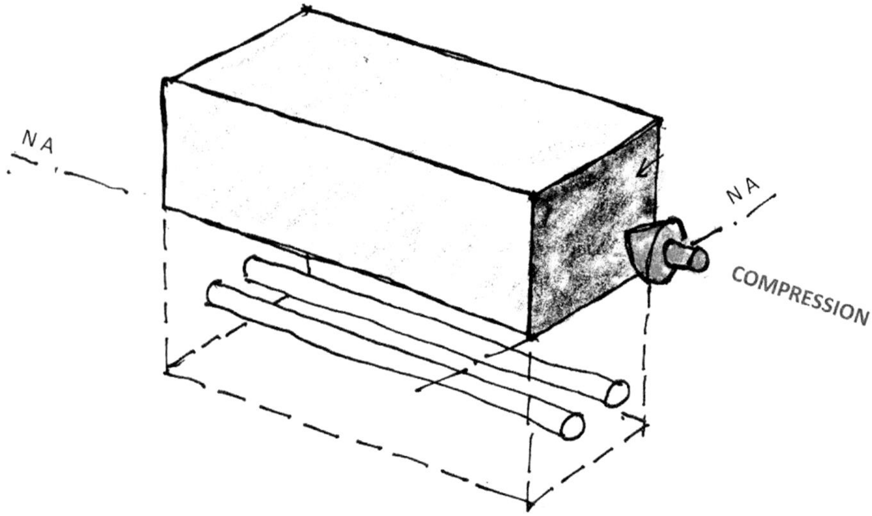

Figure 10.13a Compressive Force on Concrete Above Neutral Axis

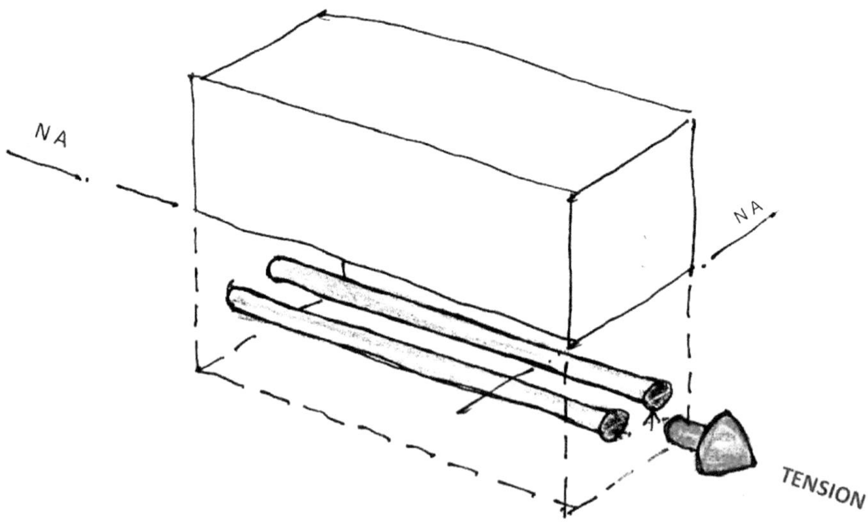

Figure 10.13b Tensile Force on Steel Below Neutral Axis

The general approach to designing a reinforced concrete beam is to assume cross-sectional dimensions, and then determine the appropriate steel reinforcement to resist moment and shear. The ACI provides recommendations for the initial assumptions of cross-section dimensions.

The Cross Section of a Typical Reinforced Concrete Beam (Figure 10.14)

Effective Depth

The effective depth is the distance (d) between the top of concrete and the centroid of reinforcing steel.

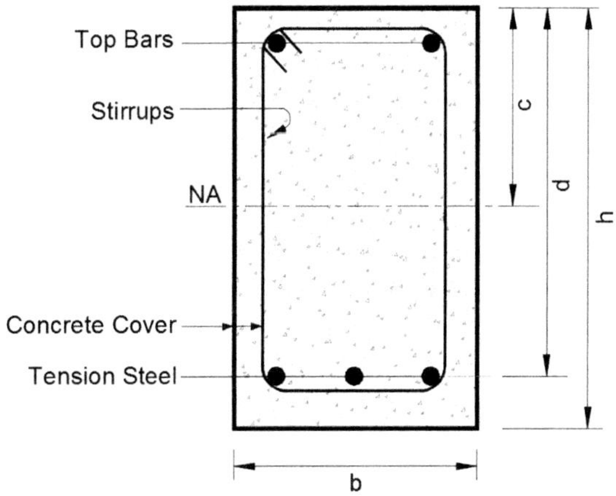

Figure 10.14 Typical Cross Section of a Reinforced Concrete Beam

Neutral Axis

The neutral axis, at a distance (c) below the top of concrete, is a function of the areas of concrete in compression and steel in tension, and is determined by ACI formulae.

Effective Area of Concrete in Compression

Although the entire area of concrete above the neutral axis is in compression, only a portion of this area is considered effective in resisting compression, as we'll see later in this chapter.

Singly and Doubly Reinforced Beams

If steel is provided only in the tensile zone, the beam is considered *singly reinforced*. If steel is also provided in the compression zone to supplement the compressive capacity of the concrete, the beam is considered *doubly reinforced*.

When singly reinforced, a beam will still have longitudinal rebar in the compression zone for the purpose of securing stirrups in a standing position. These are referred to as 'top bars' and are generally of a smaller diameter. Any additional strength in the compression zone resulting from these top bars is ignored in the design of the section. Figure 10.15 shows 'steel cage' reinforcing in a concrete beam.

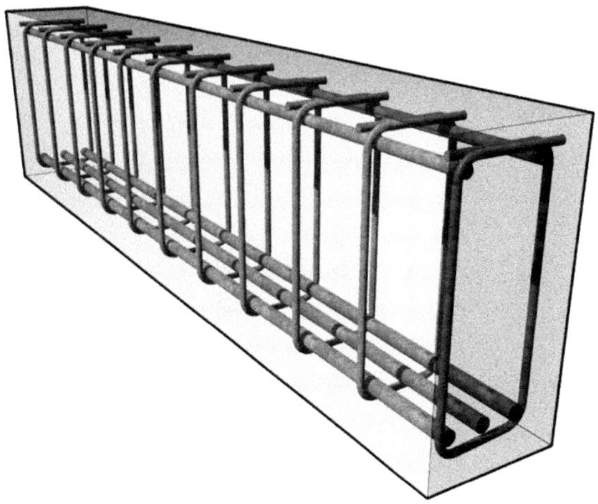

Figure 10.15 Steel Reinforcement in a Concrete Beam

Concrete Cover

The thickness of the concrete from the outside face of the steel to the exterior faces of the beam, is referred to as *concrete cover*. The cover protects the steel reinforcing from fire and corrosion, and ensures that it is sufficiently embedded to bond to the concrete and prevent slippage. The typical concrete cover for beams and columns not exposed to the weather is 1.5″.

Beam Theory

When a beam bends under load, the moment created by external forces is resisted by the beam generating an internal resisting moment. This concept applies to any beam but is especially useful for understanding reinforced concrete beams.

1. For the simple beam in Figure 10.16, the external clockwise moment created by R_1 at any distance 'a' is:

   $$\text{Moment} = \text{Force} \times \text{Distance}$$
   $$\text{Moment (at section x-x)} = R_1 \times a$$

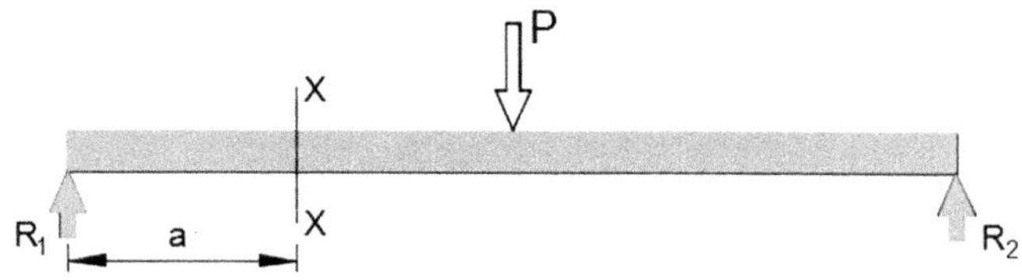

Figure 10.16 External Moment on a Beam

2. The external moment creates internal compressive and tensile stresses on the beam's cross section as shown in Figure 10.17. Figure 10.18 shows the internal stresses in terms of internal compressive and tensile forces C and T.

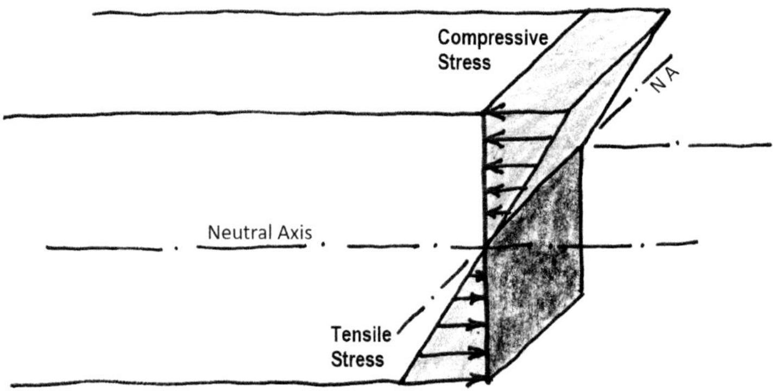

Figure 10.17 Internal Stress Distribution

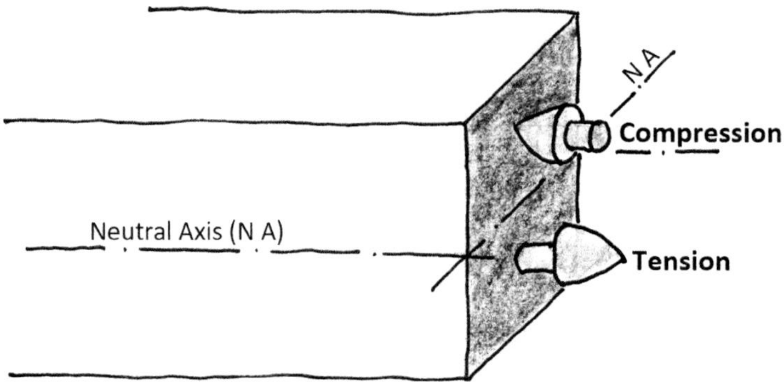

Figure 10.18 Internal Forces

3. As shown in Figure 10.19, C and T form an internal force couple creating a counter-clockwise internal resisting moment equal to:

$$(C \times y) + (T \times y)$$

Since C = T, this can be written as:

$$(C \times 2y) \text{ or } (T \times 2y)$$

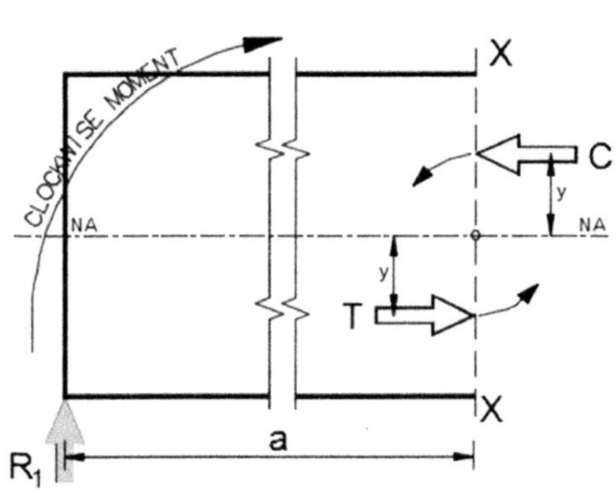

Figure 10.19 Internal Resisting Moment

Simply put, the external clockwise moment ($R_1 \times a$) generates an opposing internal resisting counter-clockwise moment equal to ($C \times 2y$) or ($T \times 2y$) (Figure 10.20).

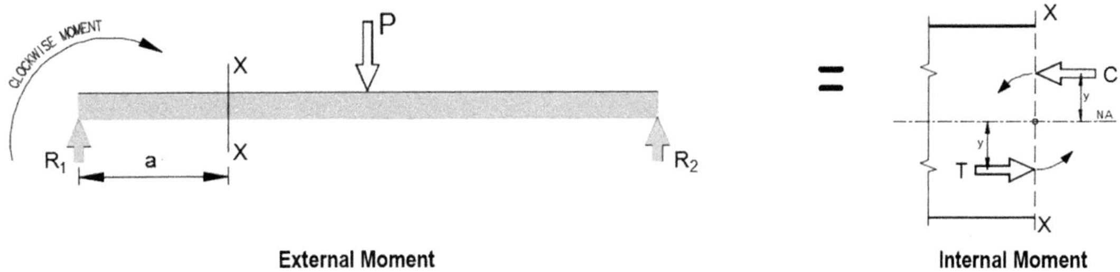

Figure 10.20 Opposing External and Internal Moments in a Beam

Internal Moment in a Reinforced Concrete Beam

Let's now examine internal moment in a reinforced concrete beam.

When a beam bends, the maximum compressive stress (f'_c) in the concrete occurs when the strain reaches 0.002. This stress reduces as the strain in concrete increases (Figure 10.21).

Since the ACI criterion for design of reinforced concrete sections is based on the strain in concrete being 0.003, the compressive stress distribution pattern is a parabolic-like volumetric shape having width b, depth c, and varying f'_c as shown in Figure 10.22.

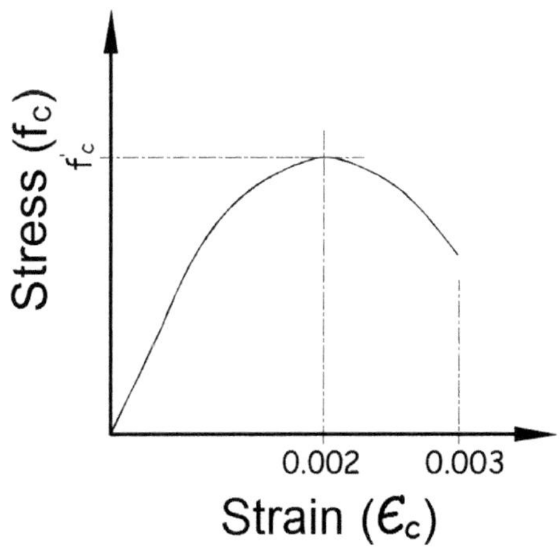

Figure 10.21 Stress Strain Diagram for Concrete in Compression

The total area of concrete in compression above the neutral axis (A_c) is ($b \times c$).

The total area of steel in tension below the neutral axis is (A_s). With the steel reaching its yield stress, the tensile stress distribution pattern is a volumetric shape of area A_s and stress of f_y.

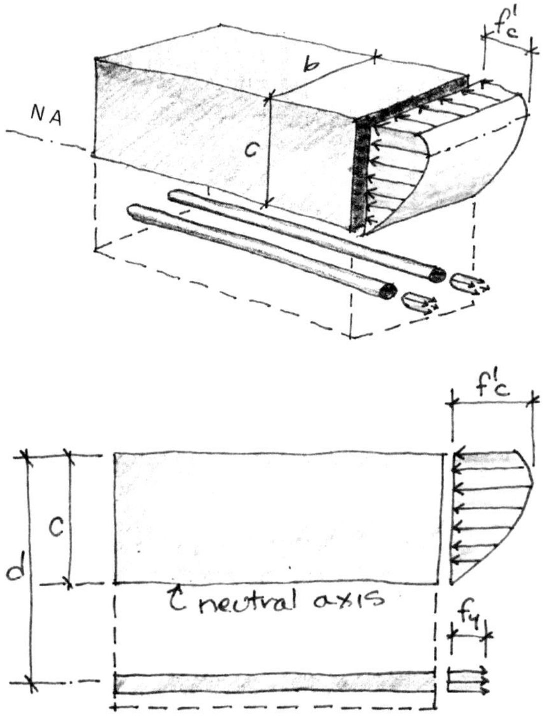

Figure 10.22 Concrete / Steel Stress Distribution Pattern

Determining the Compressive Force C

Since the total compressive stress in the volume of the parabolic shape is somewhat difficult to calculate, for simplicity it is converted into an equivalent rectangular prism known as the Whitney Stress Distribution block (Whitney block) having depth a, width b, and uniform stress $0.85f'_c$ (Figure 10.23).

Note that the *effective area* of concrete in compression, termed (A_c), is reduced to ($a \times b$).

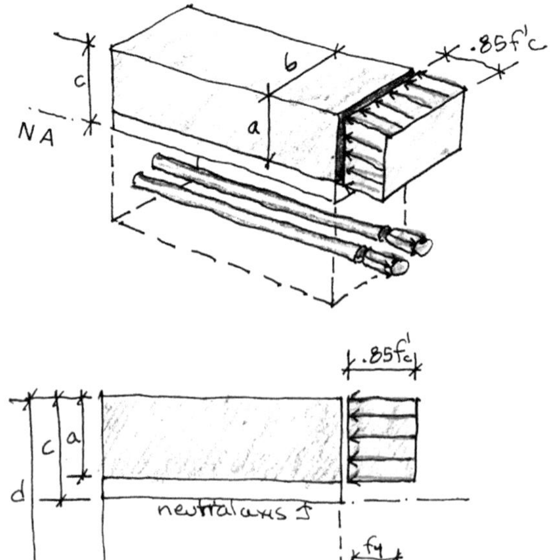

Figure 10.23 Whitney Block

If we replace the volume of Whitney block with an equivalent compressive force C, this force will act at the centroid of the Whitney block, or a/2 from the top of the section and will be (Figure 10.24):

$$\mathbf{C} = (\mathbf{0.85\ f'_c}) \times (\mathbf{A_c})$$

$$\text{since } A_c = (a \times b)$$

$$\mathbf{C} = (\mathbf{0.85\ f'_c}) \times (\mathbf{a} \times \mathbf{b})$$

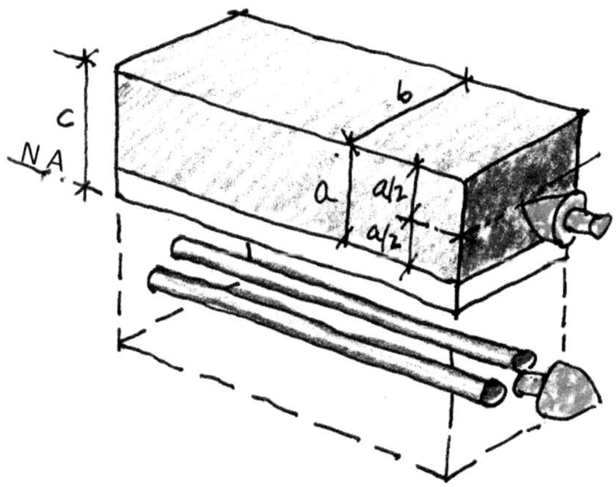

Figure 10.24 Equivalent Compressive and Tensile Force on Whitney Block

Determining the Tensile Force (T)

If we replace the volume of the tensile stress distribution pattern with an equivalent tensile force T, this force will act at the centroid of the reinforcing steel below the neutral axis and will be (Figure 10.24):

$$\mathbf{T} = (\mathbf{f_y}) \times (\mathbf{A_s})$$

Determining the Depth (a) of the Whitney Block

As an internal force couple in equilibrium, C and T must be equal. By equating them, we can solve for the depth of the Whitney block 'a' (i.e., the effective area of concrete in compression) (Figure 10.25). The ACI defines 'a' as ($\beta_1 \times c$) where:

β_1 = A flexure coefficient whose value depends on the strength of concrete
(0.85 for concrete compressive strengths of 2,500 to 4,000 psi; varies for other strengths)

c = distance of outermost compression fiber to neutral axis

$$\mathbf{C} = \mathbf{T}$$

$$(0.85\ f'_c) \times (a \times b) = (f_y) \times (A_s)$$

solving for (a):

$$a = \frac{(A_s \times f_y)}{(0.85 \times f'_c \times b)}$$

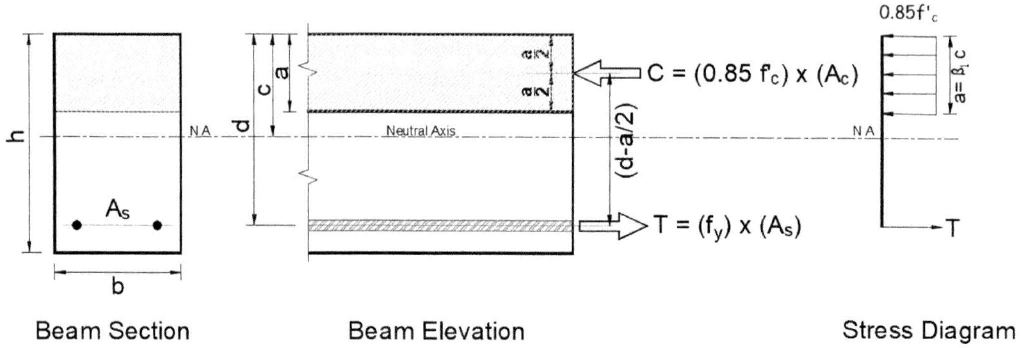

Figure 10.25 Internal Force Couple in a Beam

Tension Controlled Sections

A reinforced concrete beam can fail either by crushing of the concrete or by yielding of the steel.

- Crushing of the concrete is assumed to occur when the strain in the concrete (ϵ_c) reaches 0.003.
- Yielding of the steel starts when its stress reaches the yield point (f_y) corresponding to a strain in the steel of about 0.002.

If these two conditions occur simultaneously, it is called *balanced design*.

- If the steel provided is more than required for a balanced design, the section is called *over-reinforced* or *compression controlled*. This is undesirable since the failure of a beam would occur by the crushing of concrete and would be a sudden and catastrophic failure.
- If the steel provided is less than required for a balanced design, the section is called *under-reinforced*. If under-reinforced, the steel would reach its maximum strain first and would yield before the concrete. Such a section is termed *tension controlled* and is the preferred condition. To assure this, the ACI limits the maximum amount of steel to that which can develop a strain of 0.005 or greater (Figure 10.26).

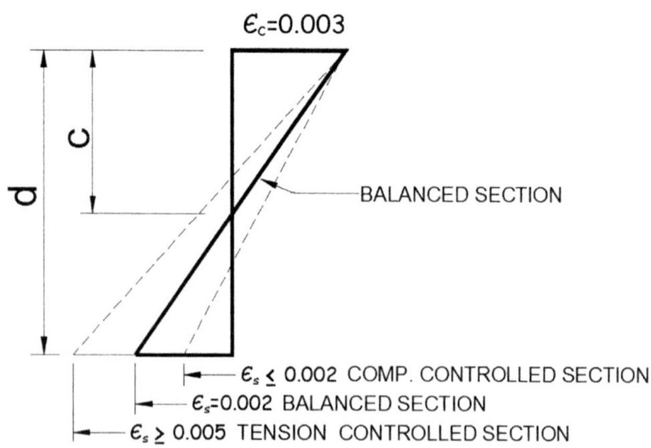

Figure 10.26 Comparative Strain Diagrams

Moment Capacity of the Beam

Since the total compression and total tension are equal, the nominal moment capacity of the beam is either C or T multiplied by the distance between them.

$$\text{Moment} = \text{Force} \times \text{Distance}$$

$$M_n = T \times (d - a/2)$$

$$\text{since } T = f_y \times A_s:$$

$$\mathbf{M_n = (f_y \times A_s) \times (d - a/2)}$$

where:
M_n is the nominal moment capacity (i.e., nominal flexural strength) of the beam.

The ACI applies the strength reduction factor (Φ) to modify various strengths. When used in flexure, Φ has a value of 0.9. Applying the strength reduction factor to M_n:

$$\Phi M_n = \Phi\ [(f_y \times A_s) \times (d - a/2)]$$

substituting for 'a':

$$\Phi M_n = \Phi A_s f_y \left\{ d - 0.59 \left[\frac{A_s \times f_y}{f'_c \times b} \right] \right\}$$

where:
ΦM_n is the available moment capacity (i.e., available flexural strength) of the beam.

Required Area of Tensile Steel

Using the above equation to determine the appropriate area of steel (A_s) is somewhat cumbersome. The ACI provides tables that simplify the selection of steel by relating the percentage of steel to the available flexural strength of the beam. One such table (1.6—Flexure Design Aids—for use with f_y of 60,000 psi steel, and f'_c varying from 3,000 to 6,000 psi concrete) is reproduced in Appendix 3.

This table uses a factor termed (ΦK_n) that is a function of the concrete compressive strength (f'_c), steel yield strength (f_y), and the steel ratio (ρ). It is based on the general strength equation $\Phi S_n \geq U$ (see Chapter 2). When applicable to flexure, this equation is expressed as:

$$\mathbf{\Phi M_n \geq M_u}$$

(i.e., the available flexural strength is equal to or greater than the required flexural strength)

where:
$\Phi M_n = \Phi K_n bd^2 / 12{,}000$
substituting for ΦM_n:
$\Phi K_n bd^2 / 12{,}000 \geq M_u$
or
$\Phi K_n \geq (M_u \times 12{,}000)/bd^2$

Based on the calculations for (ΦK_n), the percentage of steel required (ρ) is read off Table 1.6 directly.

For tension controlled sections, the ACI provides a value of 0.9 for the strength reduction factor Φ.

Designing a Reinforced Concrete Beam to Resist Maximum Moment

Assume Beam Cross-Sectional Dimensions

Determine Loads

Determine Maximum Moment and Shear

Determine Required Area of Steel (A_s)

- calculate (Φ K_n)
- from Table 1.6 in Appendix 3, determine ρ (the percentage of area of steel to effective area of concrete) corresponding to (Φ K_n)
- determine the required area of steel ($A_s = \rho \times b \times d$)

Select Size and Quantity of Rebar

From the required area of steel, select the size and quantity of rebar from available tables.

Check Limits for Reinforcement

The ACI provides the following guidelines to assure that the amount of steel is in accordance with minimum and maximum limits, and that the rebar spacing allows for the proper flow of concrete.

Strain in Steel

To assure a tension controlled section, verify the strain in steel is 0.005 or greater. This is determined from Table 1.6 using the percentage of steel provided (ρ).

Minimum Area of Steel

Too low an area of steel results in excessive cracking. The minimum percentage of area of steel to effective area of concrete is termed (ρ_{min}) and is given by:

$$\rho_{min} = (\text{minimum area of steel})/(\text{effective area of concrete}) = \left(3\ \sqrt{f'_c}\right)/f_y$$

$$(\text{but not less than } 200/f_y)$$

In terms of the area of steel;

$$\text{Minimum Area of Steel} = \rho_{min} \times (b \times d) = \left[\left(3\ \sqrt{f'_c}\right)/f_y\right] \times (b \times d)$$

$$\left[\text{but not less than } (200/f_y) \times (b \times d)\right]$$

Minimum % of Steel (ρ_{min})	Minimum Area of Steel
$(3\ \sqrt{f'_c})/f_y$	$[(3\ \sqrt{f'_c})/f_y] \times (b \times d)$
but not less than: $200/f_y$	but not less than: $(200/f_y) \times (b \times d)$

Minimum Clearances for Steel

The minimum clear rebar spacing shall be the greatest of:

- 1″
- one bar diameter
- 1.33 times the maximum aggregate size

With minimum rebar spacing established, verify that steel reinforcing properly fits within assumed section.

Resisting Shear

Shear failures in concrete beams can be catastrophic—even more so than flexural failures.

Shear Reinforcement

In a simple beam under gravity loads, the combination of flexural and shear stresses create diagonal tension cracks, generally more severe and inclined towards the ends, and more vertical but less severe towards the middle (Figure 10.27).

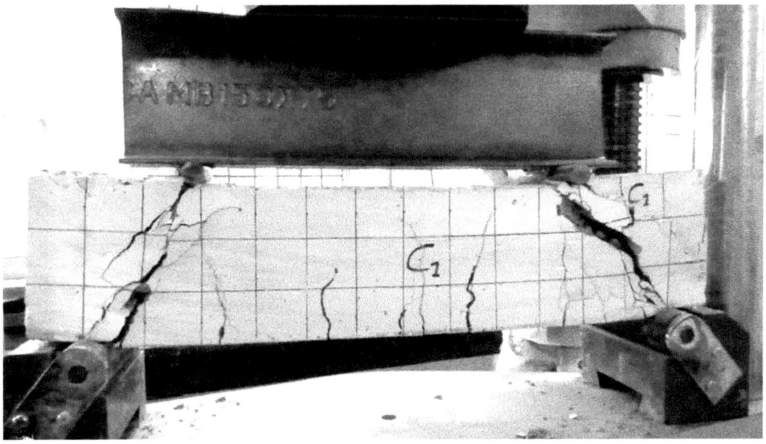

Figure 10.27 Shear Stresses in a Simply Supported Concrete Beam

The tendency of the diagonal tension stresses (flexure and shear) is to pull the cracks apart (Figure 10.28).

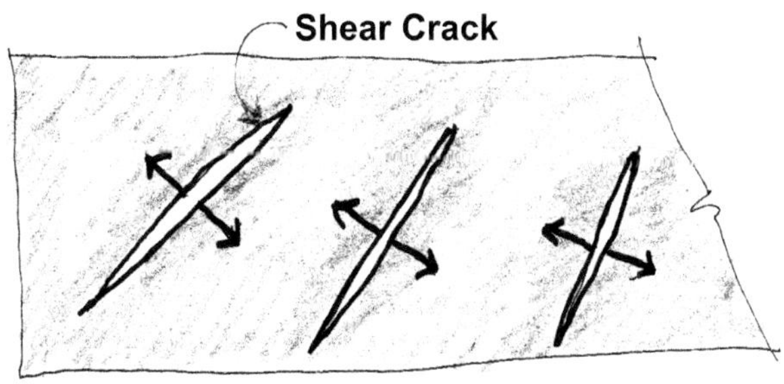

Figure 10.28 Forces Tending to Separate a Shear Crack

The most efficient way to resist the separation of the cracks would be to provide reinforcing that crosses the cracks at right angles, in a sense 'stitching' the crack together as a doctor might stitch a cut (Figure 10.29).

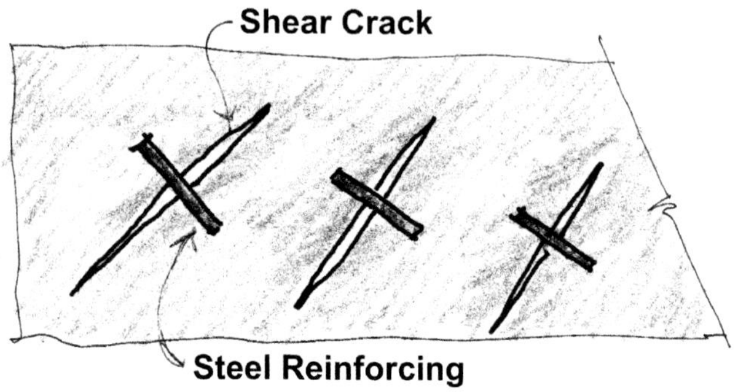

Figure 10.29 Ideal Placement of Reinforcing to Resist Crack Separation

Similarly, the ideal way to place steel reinforcing would be as shown in Figure 10.30.

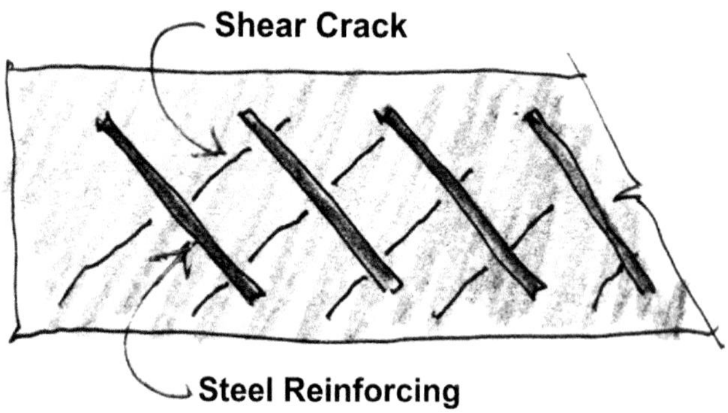

Figure 10.30 Ideal Placement of Steel Shear Reinforcing

While inclined reinforcing is possible, and the ACI has provisions for this, the more practical and widely used practice is to place shear reinforcing vertically (Figure 10.31).

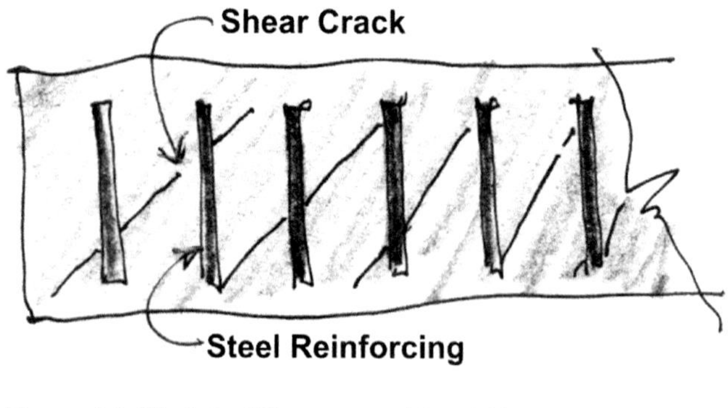

Figure 10.31 Actual Placement of Steel Shear Reinforcing

Stirrups, the term used for steel shear reinforcing, are rebar (generally #3 or #4) bent into loops and held in place by the top bars (Figure 10.32).

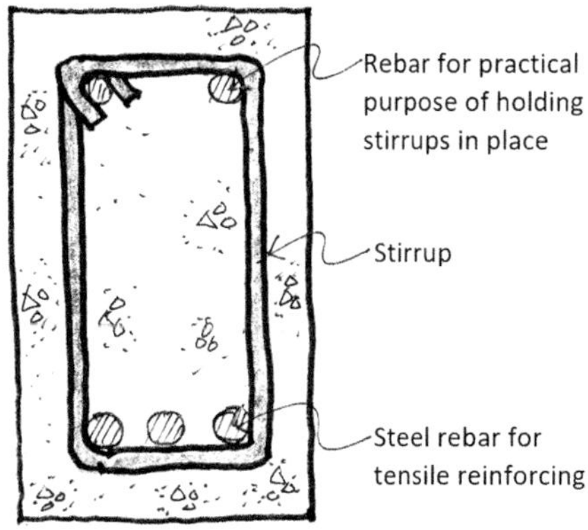

Figure 10.32 Stirrup Placement in a Concrete Beam

The Shear Capacity of a Reinforced Concrete Beam (V_n)

When applicable to shear, the general strength equation $\Phi\, S_n \geq S_u$ (see Chapter 2) is expressed as:

$$\Phi\, \mathbf{V_n} \geq \mathbf{V_u}$$

(i.e., the available shear strength is equal to or greater than the required shear strength)

Put in terms of V_n:

$$\mathbf{V_n} \geq \mathbf{V_u} / \Phi$$

When used in shear, the strength reduction factor Φ has a value of 0.75. Therefore:

$$\mathbf{V_n} \geq \mathbf{V_u} / \mathbf{0.75}$$

In a reinforced concrete beam, V_n is the sum of the nominal shear strength of the concrete (V_c) and the nominal shear strength of the steel stirrups (V_s) (Figure 10.33):

$$\mathbf{V_n} = \mathbf{V_c} + \mathbf{V_s}$$

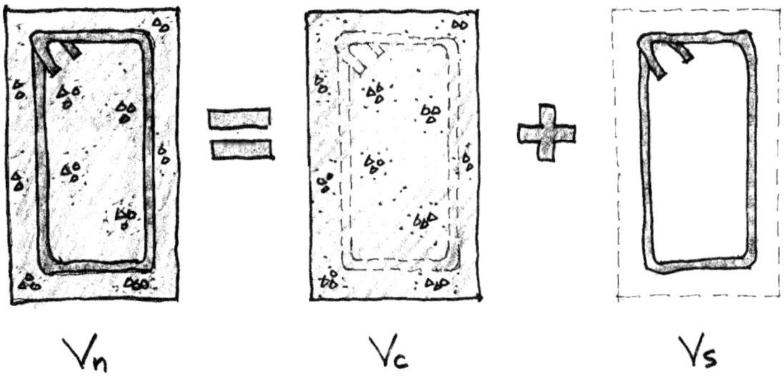

Figure 10.33 Shear Strength of a Reinforced Concrete Beam

Determine Shear Strength of Concrete Section (V_c)

The shear strength of the concrete is dependent upon its compressive strength (f'_c) and effective cross-sectional area ($b \times d$) and is given by:

$$\mathbf{V_c = 2\sqrt{f'_c} \times b \times d}$$

Determine Shear Force to be Resisted by Steel (V_s)

$$V_s = V_n - V_c$$

since $V_n = V_u/0.75$:

$$\mathbf{V_s = (V_u / 0.75) - V_c}$$

where:
V_u = required shear strength

In other words, V_s is the nominal shear force for which the steel stirrups must be designed. Note however that, with certain exceptions, the code allows V_u to be reduced to that at a distance equal to the effective depth (d) from the face of support. We'll see how this is done in Chapter 11.

Determine Stirrup Size and Spacing (s)

Although stirrups are almost always required on a typical beam, there may be situations where they are not. The ACI requires stirrups if:

$$\mathbf{V_u > 0.5\, \Phi\, V_c}$$

With V_s known, the typical approach is to assume a stirrup rebar size and then calculate its spacing and quantity. The ACI provides the following formula for calculating stirrup spacing (Figure 10.34):

$$\mathbf{s = (A_v \times f_y \times d) \;/\; V_s}$$

where:
- s = stirrup spacing (3″ to 4″ is considered the minimum practical spacing for stirrups)
- A_v = total cross-sectional area of the stirrup
- f_y = yield stress of steel
- d = effective depth of the beam
- V_s = shear force to be resisted by the steel

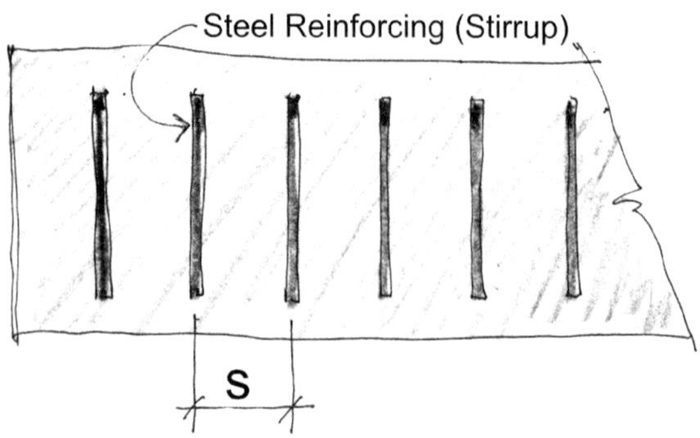

Figure 10.34 Stirrup Spacing

Check Limits for Reinforcement

- Verify that the shear capacity of steel is no greater than:

$$V_s < 4\sqrt{f'_c} \times b \times d \qquad \text{(if it is, the ACI provides additional spacing criteria)}$$

- Maximum stirrup spacing shall be the smallest of:

$$s_{max} = (A_v \times f_y)/(50 \times b)$$
$$s_{max} = d/2$$
$$s_{max} = 24$$

- Verify minimum area of stirrup is in accordance with:

$$A_{v\text{-}min} = 0.75\sqrt{f'_c} \times (\frac{b \times s}{f_y}) \qquad \text{(but not less than } 50\frac{b \times s}{f_y}\text{)}$$

Flexural and Shear Cracks

Cracks in concrete are a complex topic, but for simplicity we can consider flexural cracks to occur in regions of maximum moment, and shear cracks to occur in regions of maximum shear (near supports). These cracks are small, a normal occurrence, and not typically visible to the naked eye. Figure 10.35 shows the general locations of flexural and shear cracks in a continuous concrete beam with overhangs on each end.

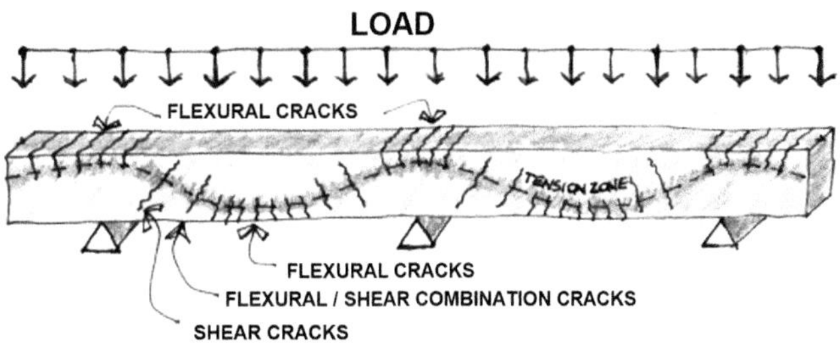

Figure 10.35 Flexural and Shear Cracks in a Reinforced Concrete Beam

Deflection

In a simply supported beam, the maximum moment (and flexural cracking) occur at the center of the span, with reducing moment and flexural cracking towards the supports. Since the amount of flexural cracking affects the beam's cross section, it also affects the beam section's moment of inertia, which will vary accordingly.

Deflection formulae depend in part upon moment of inertia—which must be evaluated in terms of a 'cracked section', taking into account effective moment of inertia. The ACI provides formulae to calculate the moment of inertia of a cracked section. These formulae are routinely used by engineers, but are beyond the scope of this text. However, based upon the beam's span, the ACI provides recommendations for beam proportions and minimum depth that take flexural cracking into account and result in beam deflections being kept within acceptable limits for typical situations.

10.4 DESIGN CONSIDERATIONS FOR COLUMNS

Column Reinforcing

Although concrete columns can be poured into virtually any shape, the two most common shapes are rectangular and circular. The selection depends primarily on compatibility with the structural system being used (such as beam and girder, flat plate, flat slab, waffle slab), aesthetic expression if exposed, and relative cost.

Reinforced concrete columns are designed with:

- vertical bars that assist in resisting compression and bending
- a system of lateral ties to resist the tendency of the vertical bars to bow outwards under compression (Figure 10.36)

The lateral tie system also resists shear stresses similar to the way stirrups resist shear stresses in a beam. For rectangular columns the lateral tie system is typically a reinforcing rod bent and tied into loops. For circular columns, the lateral tie system is typically a helical coil (Figures 10.37 and 10.38).

Figure 10.36 Outward Buckling of Vertical Reinforcing

Figure 10.37 Rectangular Column Reinforcing

Figure 10.38 Circular Column Reinforcing

The ACI provides the following recommendations for column reinforcing:

- The clear distance between vertical bars shall be at least 1.5 × bar diameter, but not less than 1.5″.
- The minimum clear cover over the outermost surface of lateral tie system shall be 1.5″.

Rectangular Columns (Figure 10.39)		
Vertical Reinforcing	**Lateral Tie System**	
Provide minimum of 4 rebar	Tie size to be: ▪ #3 if vertical rebars < # 10 ▪ #4 if vertical rebars ≥ # 10	Tie spacing is to be the minimum of: ▪ 48 × tie diameter ▪ 16 × size of vertical rebar ▪ least dimension of column

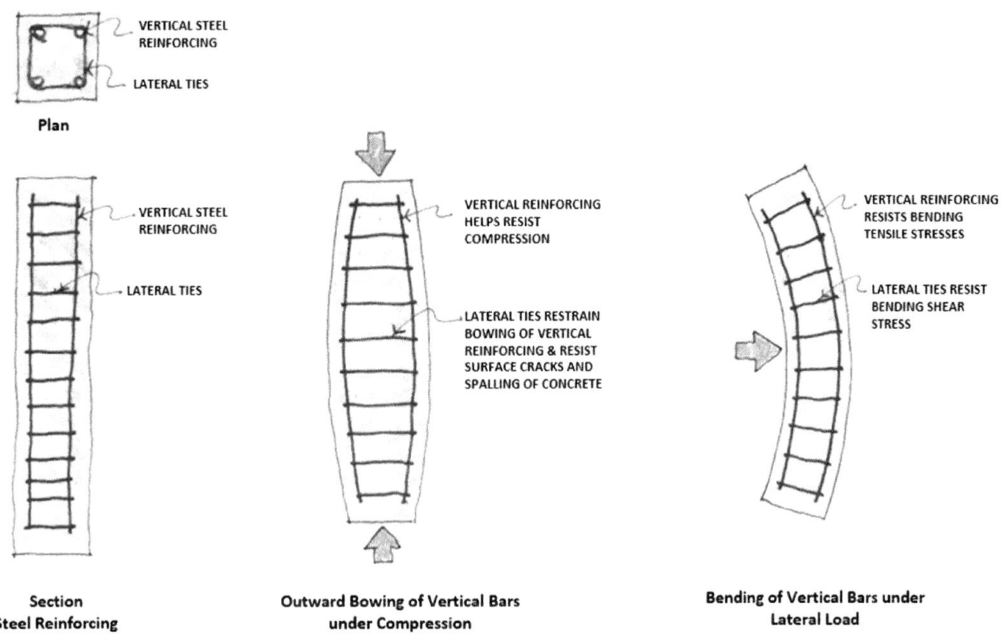

Figure 10.39 Steel Reinforcing in a Rectangular Concrete Column

Circular Columns (Figure 10.40)		
	Vertical Reinforcing	**Lateral Tie System**
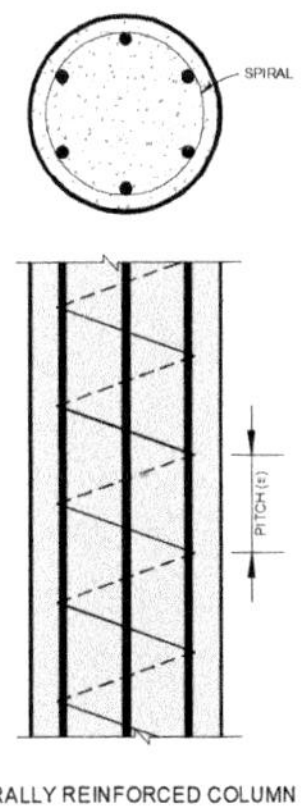 **Figure 10.40** Steel Reinforcing in a Circular Concrete Column	▪ Provide minimum of six rebars	Tie spacing is to be the minimum of: ▪ Minimum spiral tie size to be #3 ▪ Pitch (clear distance between spirals) may not exceed 3″, nor be less than 1″

Column Designation

The ACI separates columns into two categories based on whether slenderness effects (i.e., buckling) need be considered, or whether they may be neglected. For our purposes, we'll refer to columns in these two categories as 'tall' and 'short' respectively.

A column's category depends on both a) its slenderness ratio and b) whether it is part of a lateral force resisting system that allows some degree of lateral movement (referred to as sidesway), or not. Lateral force resisting systems such as shear walls and braced frames provide significant lateral resistance and are considered 'braced against sidesway'. Moment frames are much more flexible and are considered 'not braced against sidesway'.

Slenderness effects may be neglected (i.e., the column may be considered short) in the following cases:

- for compression members braced against sidesway: $\mathbf{kl_u / r \leq 40}$
- for compression members not braced against sidesway: $\mathbf{kl_u / r \leq 22}$

Tall columns (i.e., with slenderness ratios greater than the above values) are subject to additional moments, referred to as P-Delta moments, that must be considered in their design. The Reader is referred to more advanced texts for the design of tall columns.

For our Case Study, we'll assume columns to be braced against sidesway.

Compressive Capacity of Short Columns

A reinforced concrete column's total compressive capacity is determined by the combined compressive capacities of the concrete and reinforcing steel (Figure 10.41).

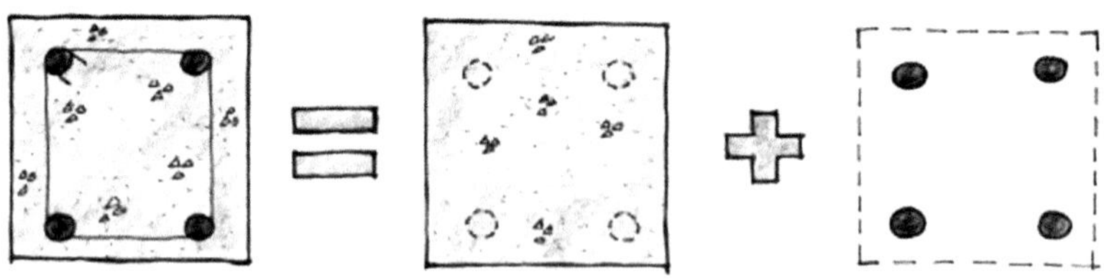

Figure 10.41 Compressive Capacity of a Reinforced Concrete Beam

The compressive capacity of the concrete and steel are a function of their compressive strengths (f'_c and f_y respectively) and their cross-sectional areas. The ratio of area of steel to area of concrete is an important design consideration having strict upper and lower limits.

For reinforced concrete columns, the general strength equation $\Phi S_n \geq U$ (see Chapter 2) is expressed as:

$\Phi \mathbf{P_n} \geq \mathbf{P_u}$ (available compressive strength ≥ required compressive strength)

Available Compressive Strength of a Short Column (ΦP_n)

The available compressive strength of a short column is given by:

For columns with tie reinforcement:

$$\Phi \mathbf{P_n} = \Phi \times \mathbf{0.80} \left[\mathbf{0.85\, f'_c} \left(\mathbf{A_g} - \mathbf{A_{st}}\right) + \left(\mathbf{A_{st}} \times \mathbf{f_y}\right)\right]$$

For columns with spiral reinforcement:

$$\Phi \mathbf{P_n} = \Phi \times \mathbf{0.85} \left[\mathbf{0.85\, f'_c} \left(\mathbf{A_g} - \mathbf{A_{st}}\right) + \left(\mathbf{A_{st}} \times \mathbf{f_y}\right)\right]$$

where:

P_n = nominal compressive strength
ΦP_n = available compressive strength
P_u = required compressive strength
Φ = strength reduction factor
0.65 for tied columns; 0.75 for spiral columns
A_g = gross area of column cross section
A_{st} = area of steel
f_y = yield strength of steel
f'_c = concrete compressive strength

Limits for the Area of Vertical Steel

The ACI establishes minimum and maximum limits for the ratio of area of reinforcing steel to area of concrete in columns. These limits are expressed as a percentage:

- minimum percentage of steel: 1% of A_g
- maximum percentage of steel: 8% of A_g

Too low a percentage of steel can lead to creep and shrinkage under sustained compressive loads. Providing a minimum percentage ensures that the column always has some resistance to bending. Too high a percentage of steel may reduce the clear space between the bars and lead to difficulty in the proper placement of concrete around the reinforcing.

Designing a Short Reinforced Concrete Column

The typical steps in the design of a short reinforced concrete column are shown in Chapter 11, Figure 11.5.

11

Design in Reinforced Concrete—Case Study

Our Case Study for design in reinforced concrete is a two-story one-way framed structure, 48 ft × 48 ft, with beams spanning 24 ft, girders spanning 16 ft, and 12-ft floor-to-floor heights. We'll focus on the structural design of the first floor and select typical members Beam 3, Girder B, and Column B2 to design. In addition to the loads from the first floor, the loads from the roof will be added onto Column B2. Figure 11.1 shows the first floor framing plan.

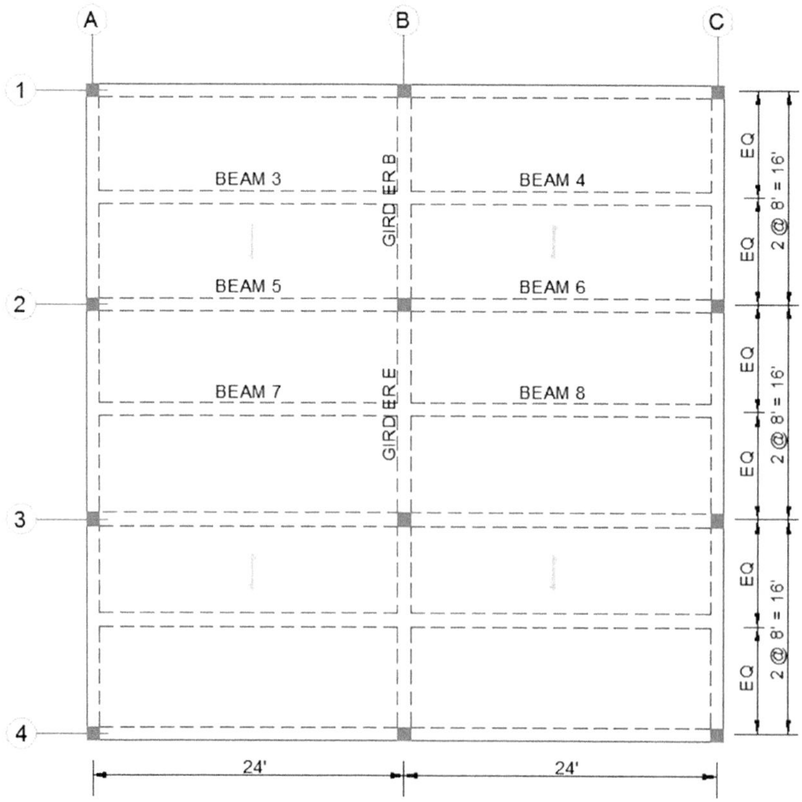

Figure 11.1 First Floor Framing Plan

We'll design the typical members using Strength Design based on ACI 318–11, Building Code Requirements for Structural Concrete. This document contains tables, charts, and guidelines to assist in the design of structural members and perform checks (see Appendix 3).

11.1 ASSUMPTIONS

Fixity of Connections

Connections between concrete members such as beams and columns are typically cast monolithically with continuous reinforcing steel extending from one member to the other (Figure 11.2). Such concrete connections develop a degree of fixity having the ability to resist moment. Although fixity of these connections does play a large part in concrete design, for simplicity in the Case Study we'll assume beams and girders to have pinned connections without the ability to resist moment. The columns will be considered to be part of a lateral force resisting system.

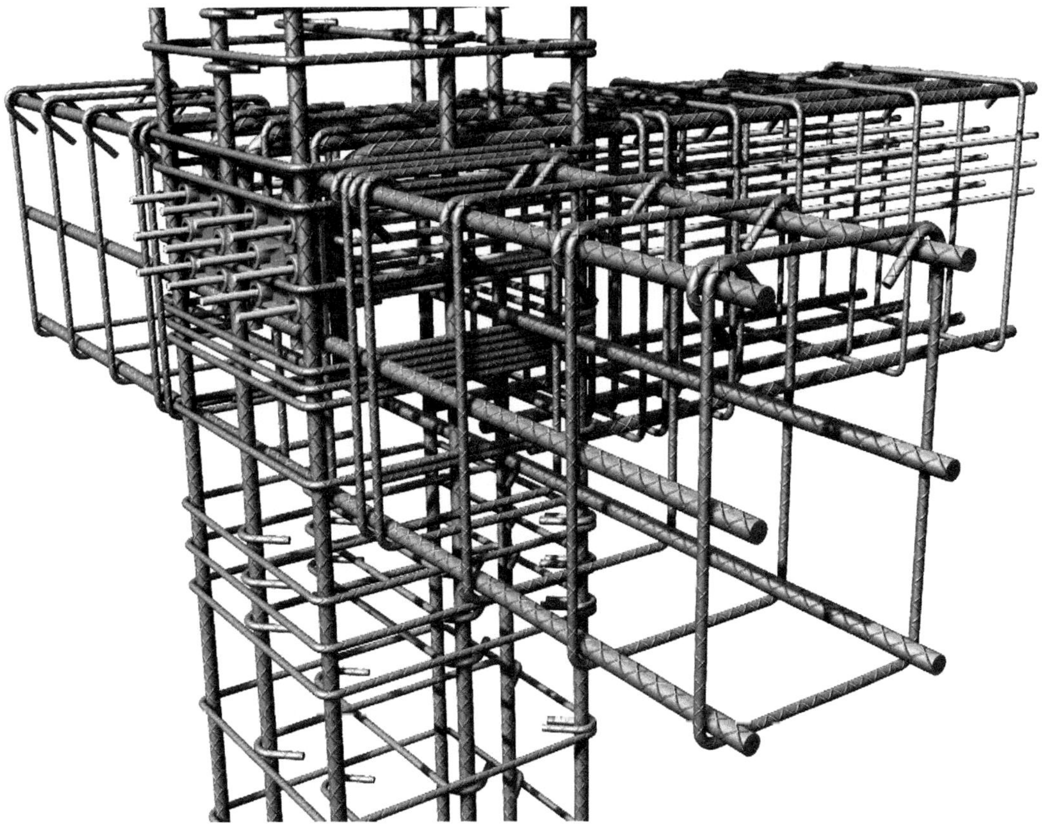

Figure 11.2 Continuous Reinforcing in Concrete Connections

Floor Construction

Floor construction will be a 5″ concrete slab spanning the concrete beams as a one-way slab.

A slab, when poured monolithically with a beam, offers the advantage of 'T-action' wherein the slab adds to the compressive strength of the beam. However, as noted above, we'll design the beam (as shown dashed) independently of the slab (Figure 11.3).

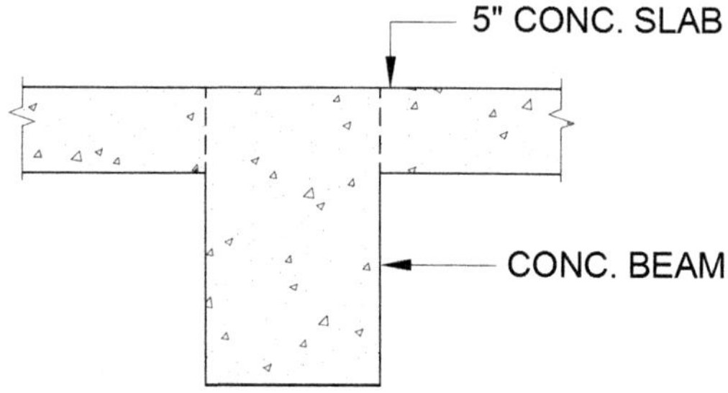

Figure 11.3 Floor Construction

Material Properties

Structural members will be rectangular-shaped with the following material properties:

- concrete compressive strength (f'_c): 4,000 psi
- tensile yield strength of reinforcing steel (f_y): 60 ksi
- modulus of elasticity (E_c): reinforced concrete's modulus of elasticity is dependent upon its grade and weight and is given by:

  $$E_c\,(\text{ksi}) = 57{,}000\ \sqrt{f'_c}\,(\text{psi})$$

  For the 4,000 psi normal weight concrete in our Case Study:

  $$E_c\,(\text{ksi}) = 57{,}000\ \sqrt{4{,}000} = 3{,}605\ \text{ksi}$$
- density of normal weight concrete: 150 lb/ft^3

Floor and Roof Loads

We'll assume the following service loads:

Floor Loads	Roof Loads
Dead Load *5″ slab @ 150 lb/ft^3 = 62.5 psf ceiling, floor finishes = 12.5 psf **Total = 75.0 psf**	***Dead Load*** *5″ slab @ 150 lb/ft^3 = 62.5 psf ceiling, floor finishes = 12.5 psf **Total = 75.0 psf**
Live Load **Total = 75.0 psf**	***Live Load*** **Total = 30.0 psf**

*In concrete design, the self-weight of structural members is relatively large in relation to the other dead loads and is added into the calculations for dead load of the individual members.

Beams and Girders

Beams and girders will be considered to have:

- simple supports
- full lateral bracing

Figure 11.4 shows the typical steps we'll follow in the design of Beam 3 and Girder B:

1. design to resist moment
2. design for shear
3. check for deflection

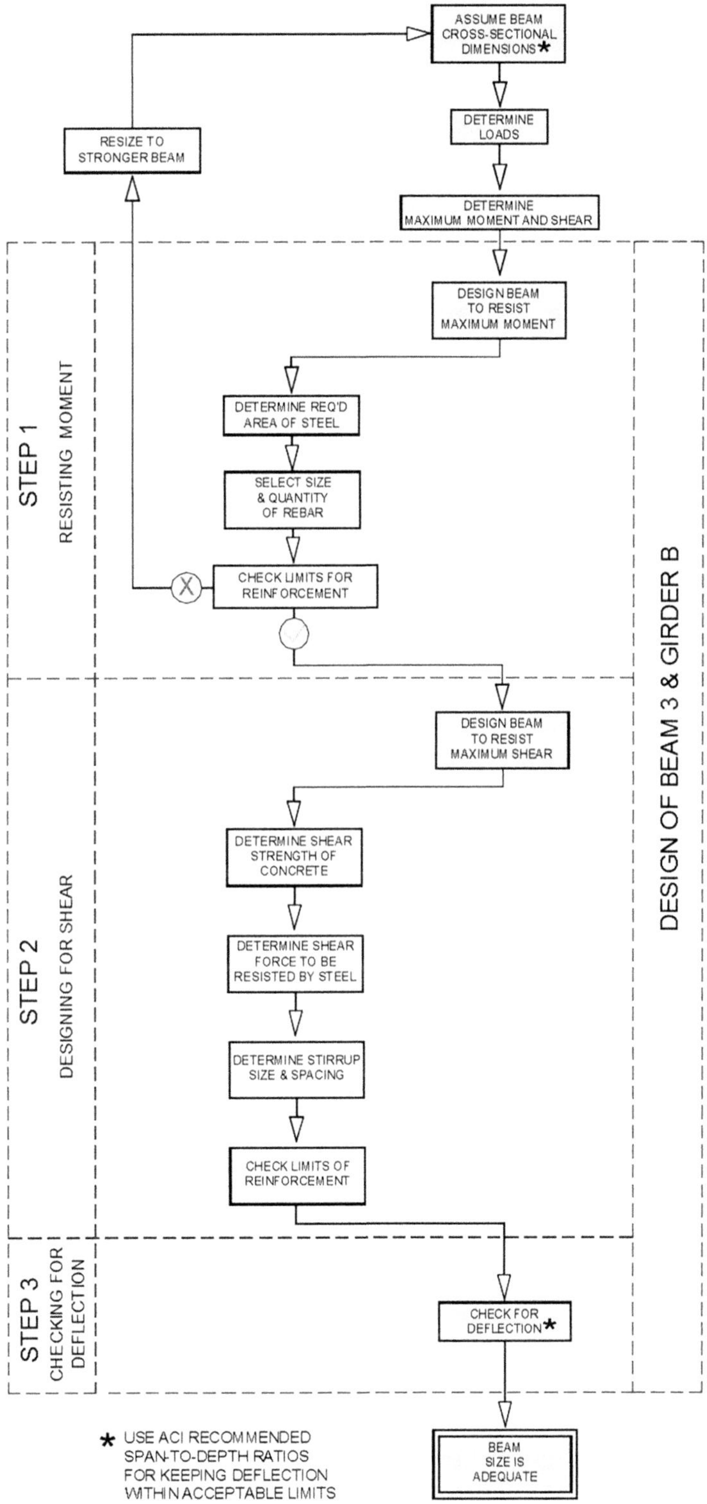

Figure 11.4 Flow Chart for Design of Beam 3 and Girder B

Columns

Columns will be considered to be part of a lateral force resisting system and have:

- axial loads only, not subject to lateral loads
- floor-to-floor height (L) = 12 ft
- pinned top and bottom restraints; therefore k-factor = 1

Figure 11.5 shows the typical steps we'll follow in the design of Column B2.

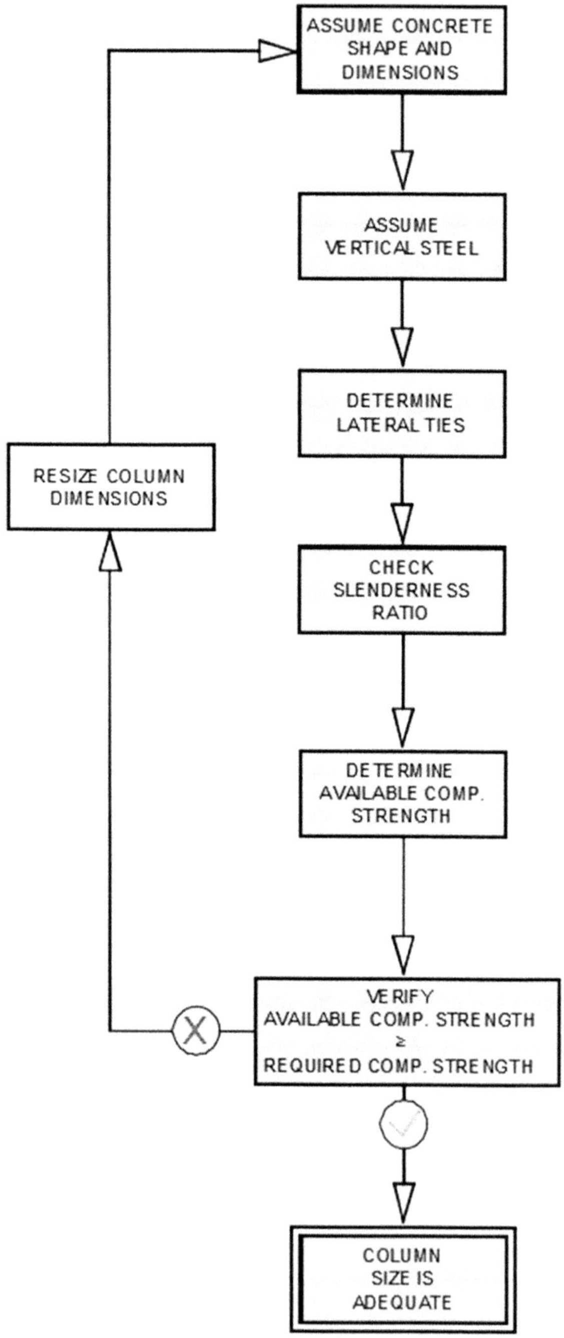

Figure 11.5 Flow Chart for Design of Column B2

Let's proceed to design our typical members based on Strength Design.

11.2 BEAM 3

Assume Beam Cross-Sectional Dimensions (Beam 3)

To determine the total load on Beam 3, its self-weight must first be determined and added onto the floor dead loads. Although experienced designers can make an educated guess for an appropriate beam depth and width, ACI 318 recommends a minimum span-to-depth ratio of L/16 for simply supported beams. With this recommendation, deflections are kept within acceptable limits and need not be checked.

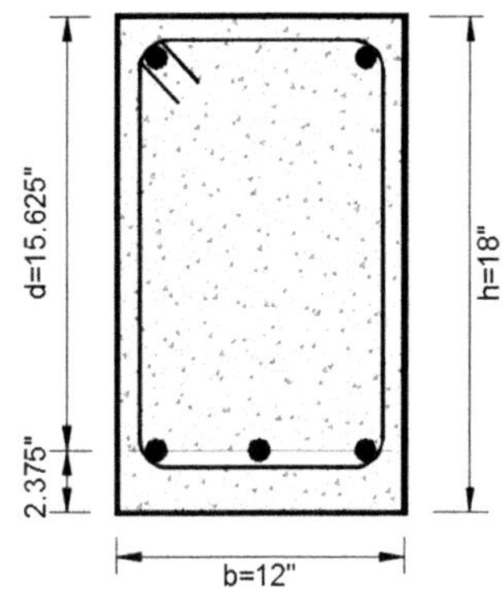

Figure 11.6 Beam 3 Cross Sectional Assumptions

Beam 3 has a span (L) of 24 ft.
The minimum assumed height (h) is therefore:

$$h = L/16 = (24 \times 12)/16 = 18 \text{ in}$$

A reasonably proportioned beam width b for this assumed depth is 2/3 the depth. The minimum assumed width is therefore:

$$b = 2/3 \times h = 2/3 \times 18 = 12 \text{ in}$$

We'll assume #8 tensile rebar (1″ diameter), #3 stirrups (0.375″ diameter), and a clear cover of 1.5″. The assumed effective depth is therefore:

$$d = 18.0 - 1.5 - 0.375 - 0.5$$
$$= 15.625 \text{ in (Figure 11.6)}$$

Determine Loads (Beam 3)

Calculate the uniform load imposed on Beam 3.

The load tributary area for Beam 3 is a section of floor 8 ft × 24 ft (Figure 11.7). The uniformly distributed loads (w) on Beam 3 are:

$$w = \text{uniform dead load from floor}$$
$$= 75 \text{ lb/ft}^2 \times 8 \text{ ft} = 600 \text{ lb/ft} \qquad = 0.60 \text{ k/ft}$$
$$w = \text{uniform dead load from self-weight of beam}$$
$$= b \times h \times \text{weight/ft}^3 \text{ (changing inches to ft, and lbs to k)}$$
$$= (12/12) \times (18/12) \times 150/1{,}000 \qquad = 0.23 \text{ k/ft}$$

$$w_{DL} = \text{uniform (total) dead load} \qquad = 0.83 \text{ k/ft}$$
$$w_{LL} = \text{uniform live load}$$
$$= 75 \text{ lb/ft}^2 \times 8 \text{ ft} = 600 \text{ lb/ft} \qquad = 0.60 \text{ k/ft}$$

$$w_{TL} = \text{uniform total (service) load} \qquad = 1.43 \text{ k/ft}$$

Applying the governing Strength Design load combination $(1.2\,D + 1.6\,L)$ to w_{DL} and w_{LL} (see Chapter 4):

Total Design Load ($\boldsymbol{w_u}$)

$$\mathbf{w_u} = 1.2\, w_{DL} + 1.6\, w_{LL}$$
$$= (1.2 \times 0.83) + (1.6 \times 0.60)$$
$$= 1.00 + 0.96$$
$$= \mathbf{1.96\ k/ft}$$

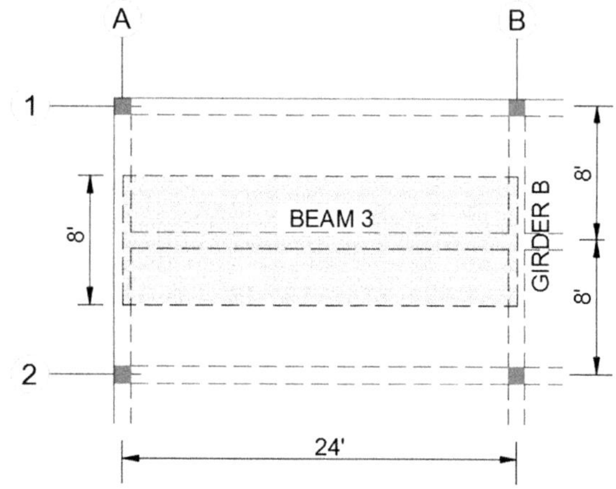

Figure 11.7 Load Tributary Area of Beam 3

Determine Reactions, Maximum Shear, Maximum Moment (Beam 3)

Beam 3 is simply supported with a uniformly distributed load along its length. From Appendix 4—Beam Diagrams and Formulae—the shear and moment diagrams with maximum values for Beam 3 are shown in Figure 11.8.

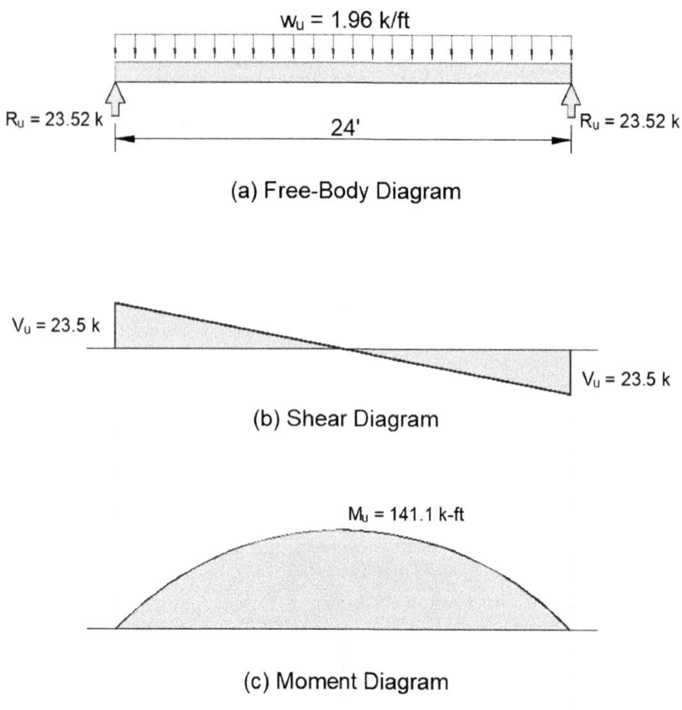

Figure 11.8 Beam 3: Free-body, Shear & Moment Diagrams

$$L\ (\text{length}) = 24\ \text{ft}$$
$$w_u = 1.96\ \text{k/ft}$$

Reactions (R_u)

$$R_u = w_u L/2$$
$$R_u = (1.96 \times 24)/2$$
$$\mathbf{R_u = 23.52\ k}$$

Maximum Shear (V_u)

$$V_u = R_u$$
$$\mathbf{V_u = 23.52\ k}$$

(Required Shear Strength)

Maximum Moment (M_u)

$$M_u = w_u L^2/8$$
$$M_u = (1.96 \times 24^2)/8$$
$$\mathbf{M_u = 141.1\ k\text{-}ft}$$

(Required Flexura Strength)

Design of Beam 3

Step 1: Resisting Moment (Beam 3)

Determine Required Area of Steel (A_s)

- Calculate (ΦK_n)

$$\begin{aligned}\Phi K_n &= (M_u \times 12{,}000)/bd^2 \\ &= (141.1 \times 12{,}000)/(12 \times 15.625^2) \\ &= 577.95 \text{ lb/in}^2\end{aligned}$$

- From Table 1.6 (Appendix 3), determine the percentage of steel (ρ) corresponding to (ΦK_n)
  $\rho = 1.18\% = 0.0118$ (by interpolation)
- Determine the required area of steel (A_s)

$$\begin{aligned}A_s &= \rho \times b \times d \\ &= 0.0118 \times 12 \times 15.625 \\ &= 2.21 \text{ in}^2\end{aligned}$$

Select Size and Quantity of Rebar

From Table 11.1 select a size and quantity of rebar that will provide a minimum A_s of 2.21 in^2.

Several options are viable:

Rebar	Area of Steel (A_s)	
(6) #6	6 × 0.44 =	2.64 in^2
(4) #7	4 × 0.60 =	2.40 in^2
(3) #8	3 × 0.79 =	2.37 in^2

Table 11.1 Rebar Properties

	Nominal Area (A_s)	Nominal Diameter
Bar Size	**(in^2)**	**(in)**
#3	0.11	0.375
#4	0.20	0.500
#5	0.31	0.625
#6	0.44	0.750
#7	0.60	0.875
#8	0.79	1.000
#9	1.00	1.128
#10	1.27	1.270
#11	1.56	1.410
#14	2.25	1.693
#18	4.00	2.257

The generally preferred arrangement of rebar is a single layer with sufficient distance between the steel for the flow of concrete. The advantage of this arrangement is to maximize the effective depth and simplify the placement of rebar in the field.

For Beam 3 we'll select (3) #8 rebars in a single layer, with an A_s of 2.37 in^2 and a rebar diameter of 1″.

Check Limits for Reinforcement

- Strain in Steel

  To assure a tension controlled section, verify the strain in steel is 0.005 or greater.

  Calculate the percentage of steel provided (ρ):

Area of Steel Provided (A_s)	Effective Area of Concrete ($b \times d$)	% of Steel Provided (ρ) = $A_s/(b \times d)$
2.37 in^2	$12 \times 15.625 = 187.5$ in^2	$2.37 \times 100/187.5 = 1.28\%$

  From Table 1.6 (Appendix 3):

  for (ρ) of 1.28%, the strain in steel = 0.0083

  Since this strain is greater than 0.005, the section is tension controlled as desired.

- Minimum Area of Steel

  The minimum area of steel is given by:

Minimum Area of Steel	
$[(3\sqrt{f'_c})/f_y] \times (b \times d)$	$= (3\sqrt{4,000} / 60,000) \times (12 \times 15.625)$ $= 0.59$ in^2
but not less than: $(200/f_y) \times (b \times d)$	$= (200 / 60,000) \times (12 \times 15.625)$ $= 0.63$ in^2

  The area of steel provided (2.37 in^2) is indeed greater than the minimum area (0.63 in^2).

- Minimum Clearances for Steel

  For (3) #8 bars, the minimum clear rebar spacing shall be the greatest of:

  ✓ 1″
  ✓ one bar diameter = 1″
  ✓ 1.33 times the maximum aggregate size (assuming 3/4″ maximum aggregate size) $1.33 \times 0.75'' = 0.997''$

  The minimum clear rebar spacing is therefore 1″

  Verify that steel reinforcing properly fits within assumed section:

  As shown in Figure 11.9, the minimum beam width is 9″, therefore our 12″ width assumption is adequate:

Clear cover	= 1.5″
Stirrup diameter	= 0.375″
Rebar size	= 1.0″
Min. clear rebar spacing	= 1.0″

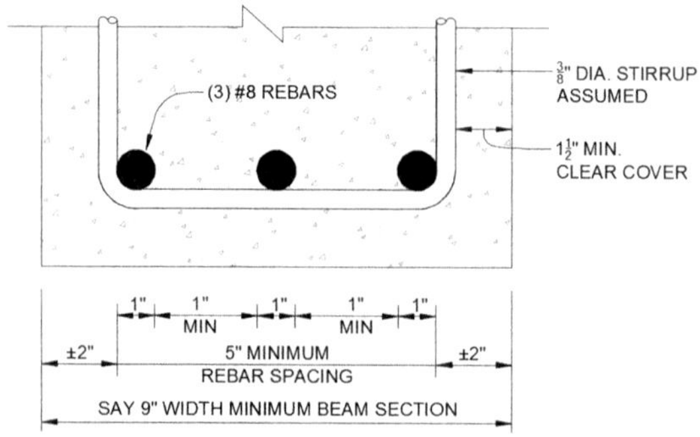

Figure 11.9 Minimum Width for Beam 3

Tables, charts, and other design aids are available from the ACI and other organizations to simplify the verification of maximum number of rebars in a beam. Using Design Aid 1–6 in Appendix 3, for the given assumptions, a 12″ wide beam could accommodate (4) #8 rebars.

Step 2: Designing for Shear (Beam 3)

Determine Shear Strength of Concrete Section (V_c)

$$\mathbf{V_c = 2\sqrt{f'_c} \times b \times d}$$
$$= [(2\sqrt{4,000}) \times 12 \times 15.625]/1,000$$
$$= 23.72 \text{ k}$$

(changing lbs to kips)

where:
- V_c = nominal shear strength of concrete
- f'_c = concrete compressive strength
- b = width of beam (12″)
- d = effective depth (15.625″)

Determine Shear Force to be Resisted by Steel (V_s)

$$\mathbf{V_s = (V_u / 0.75) - V_c}$$

where:
- V_s = nominal shear force to be resisted by steel
- V_u = maximum shear force = 23.52 k
- V_u @d = shear at distance (d) from face of support
- w_u = uniform load = 1.96 k
- L = length of beam (24 feet)

The formula for shear at a distance (d) from the face of the support is:

$$\mathbf{V_u @d = w_u (L/2 - d)}$$
$$= 1.96\ (24/2 - 15.625/12)$$
$$= 20.96 \text{ k}$$

For simplicity we'll take "d" to be the distance from the centerline of the support—a more conservative value (Figure 11.10).

Therefore:

$$V_s = (20.96/0.75) - 23.72$$
$$= 4.23 \text{ k}$$

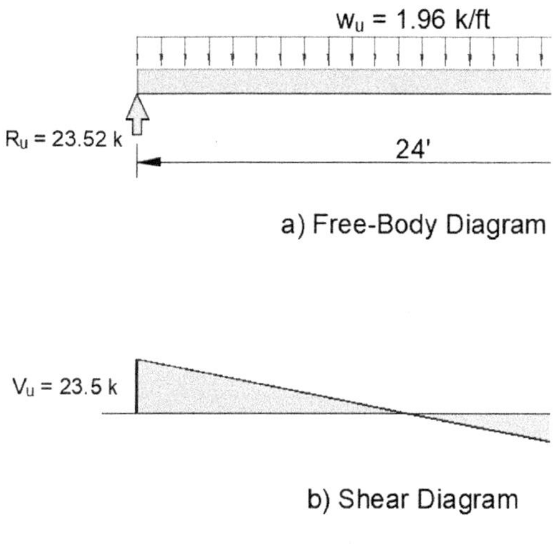

Figure 11.10 Shear at Distance 'd' From Support

Determine Stirrup Size and Spacing (s)

- Stirrups are required if:

$$V_u > 0.5\ \Phi\ V_c$$
$$> 0.5 \times 0.75 \times 23.72$$
$$> 8.9\ k$$

where: $V_u = 20.96$ k
$V_c = 23.72$ k
$\Phi = 0.75$

Since 20.96 k > 8.9 k, stirrups are required.

- Assume a stirrup rebar size (A_v)

Assume #3 with an area of 0.11 in^2 (Table 11.1).

- Calculate stirrup spacing (s)

$$\mathbf{s = (A_v \times f_y \times d) / V_s}$$

Since the stirrup is bent into a loop, the cross-sectional area of stirrup steel (A_v) is twice the cross-sectional area of the rebar, therefore: $A_v = 2 \times 0.11 = 0.22$ in^2

$$s = (0.22 \times 60 \times 15.625) / 4.23$$
$$= 206.25 / 4.23$$
$$= 48.8''\ \text{o.c.}$$

Check Limits for Reinforcement

- Verify that the shear capacity of steel is no greater than:

$$\mathbf{V_s < 4\sqrt{f'_c} \times b \times d}$$
$$(V_s) = 4.23\ k$$
$$(4\sqrt{f'_c} \times b \times d) = 4 \times \sqrt{4000} \times 12 \times 15.25/1000 = 46.30\ k$$

Since 4.23 k < 46.30 k, the section complies.

- Maximum stirrup spacing shall be the smallest of:

$$s_{max} = (A_v \times f_y)/(50 \times b) = (0.22 \times 60{,}000)/(50 \times 12) = 22''$$
$$s_{max} = d/2 = 15.625/2 = 7.8''$$
$$s_{max} = 24''$$

The smallest from the above is 7.8″ o.c.

We'll assume #3 stirrups @7″ o.c. along the entire length of the beam, with (2) #4 top rebars to hold the stirrups in place. The first stirrup is normally placed about 2″ from face of support.

- Verify minimum area of stirrup:

Minimum Area of Stirrup		
$A_{v\text{-}min} = = 0.75\sqrt{f'_c}\,\dfrac{b \times s}{f_y}$	$(0.75\sqrt{4000} \times 12 \times 7) / 60{,}000 =$	$0.085\ in^2$
but not less than: $50\,\dfrac{b \times s}{f_y}$	$(50 \times 12 \times 7) / 60{,}000 =$	$0.09\ in^2$

The area of stirrup provided ($A_v = 0.22\ in^2$) exceeds the above minimums.

Step 3: Checking for Deflection (Beam 3)

Since we've used the code-recommended span-to-depth ratios, deflections are considered to be kept within acceptable limits and need not be checked.

Figure 11.11 shows the final design for Beam 3.

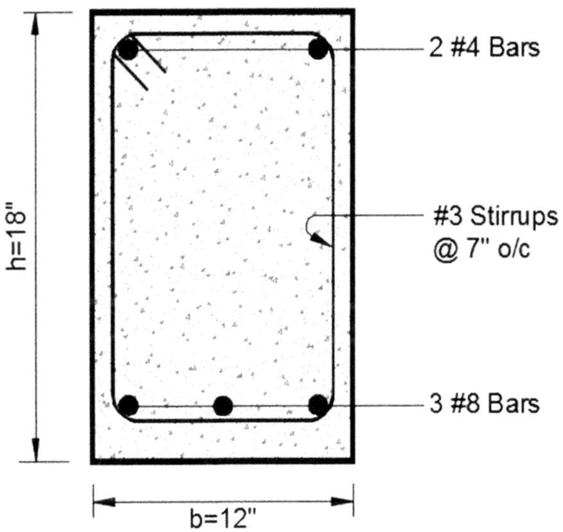

Figure 11.11 Beam 3 Final Design

11.3 GIRDER B

For the design of Girder B, we'll follow steps similar to those in Beam 3.

Assume Girder Cross-Sectional Dimensions (Girder B)

Girder B has a span (L) of 16 ft.

Although the span of Girder B is shorter than that of Beam 3, it is carrying a heavier and concentrated load at mid spans. Therefore, let's assume the girder depth to be deeper than the beam.

We'll assume the following dimensions:

$$h = 20 \text{ in}$$
$$b = 12 \text{ in}$$

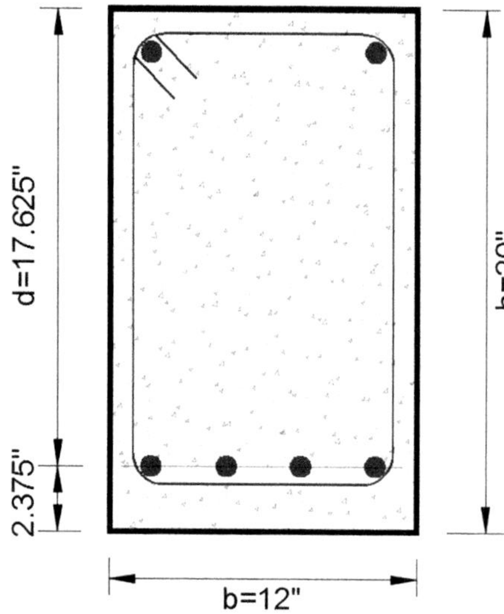

Figure 11.12 Girder B Assumptions

As in Beam 3, we'll assume #8 tensile rebar (1″ diameter), #3 stirrups (0.375″ diameter), and a clear cover of 1.5″. The assumed effective depth is therefore:

$$d = 20 - 1.5 - 0.375 - 0.5$$
$$= 17.625 \text{ in (Figure 11.12)}$$

The self-weight of Girder B is therefore:

$$= (b) \times (h) \times (\text{weight of concrete})$$
$$= (12/12) \times (20/12) \times 150/1{,}000 = 0.25 \text{ k/ft}$$
(changing inches to ft, and lbs to k)

Applying the dead load factor of 1.2:

$$\text{The factored self-weight of Girder B } (w_u) = 0.25 \text{ k/ft} \times 1.2 = 0.30 \text{ k/ft}$$

Determine Loads (Girder B)

Girder B is supporting Beam 3 and Beam 4 at its midspan. The combined design point load (P) from these two identical beams is 47.1 k (i.e., end reactions of each beam; 23.52 k + 23.52 k) (Figure 11.13).

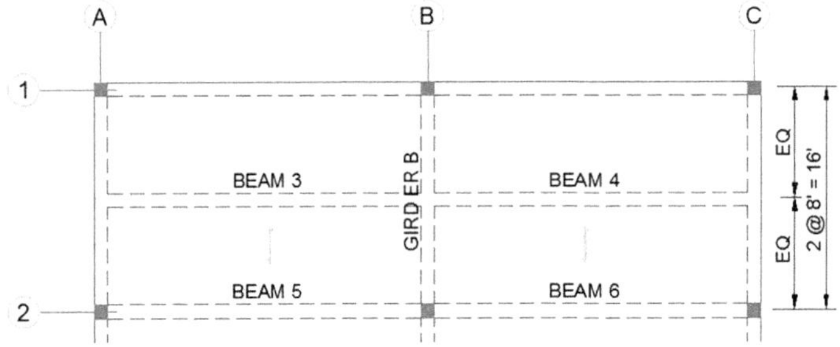

Figure 11.13 Loads on Girder B

Determine Reactions, Maximum Shear, Maximum Moment (Girder B)

Girder B is simply supported with a point load at its midspan and a uniformly distributed load along its length (i.e., its self-weight). By conventional methods the shear and moment diagrams with maximum values for Girder B are shown in Figure 11.14.

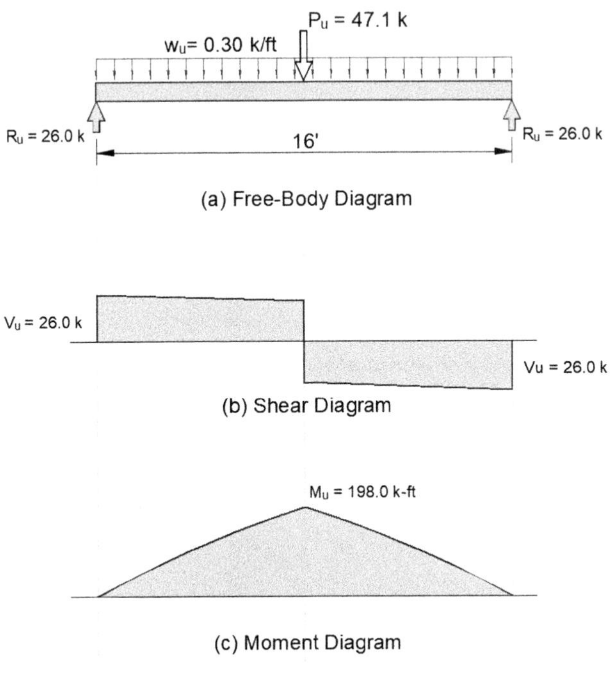

Figure 11.14 Girder B: Free-body, Shear & Moment Diagrams

$$L\ (\text{length}) = 16\text{ ft}$$
$$P_u = 47.1\text{ k}$$
$$w_u = 0.30\text{ k/ft}$$

Reactions (R_u)

$$R_u = P_u/2 + w_uL/2$$
$$R_u = (47.1\ /2) + (0.30\times 16/2)$$
$$R_u = 23.55 + 2.40$$
$$\mathbf{R_u = 26.0\ k}$$

Maximum Shear (V_u)

$$V_u = R_u$$
$$\mathbf{V_u = 26.0\ k}$$

Maximum Moment (M_u)

$$M_u = P_uL/4 + wL^2/8$$
$$M_u = (47.1\times 16/4) + (0.30\times 16^2)/8$$
$$M_u = 188.4 + 9.60$$
$$\mathbf{M_u = 198.0\ k\text{-}ft}$$

Design of Girder B

Step 1: Resisting Moment (Girder B)

Determine the Required Area of Steel (A_s)

- Calculate (Φ K_n)

$$\begin{aligned}\Phi K_n &= M_u \times 12 \times 1000/bd^2 \\ &= (198.0 \times 12 \times 1000)/(12 \times 17.625^2) \\ &= 637.39 \text{ lb/in}^2\end{aligned}$$

- From Table 1.6 (Appendix 3), determine the percentage of steel (ρ) corresponding to (Φ K_n)

$$\rho = 1.34\% = 0.0134 \quad \text{(by interpolation)}$$

- Determine the required area of steel (A_s)

$$\begin{aligned}A_s &= \rho \times b \times d \\ &= 0.0134 \times 12 \times 17.625 \\ &= 2.83 \text{ in}^2\end{aligned}$$

Select Size and Quantity of Rebar

From Table 11.1 select a size and quantity of rebar that will provide a minimum A_s of 2.83 in^2.

Several options are viable:

Rebar	Area of Steel (A_s)	
(7) #6	7 × 0.44 =	3.08 in^2
(5) #7	5 × 0.60 =	3.00 in^2
(4) #8	4 × 0.79 =	3.16 in^2

As in Beam 3, we'll choose rebars to fit in a single layer.

For Girder B, we'll select (4) #8 rebars in a single layer, with an A_s of 3.16 in^2 and a bar diameter of 1″.

Check Limits for Reinforcement

- Strain in Steel

  To assure a tension controlled section, verify the strain in steel is 0.005 or greater.

  Calculate the percentage of steel provided (ρ):

Area of Steel Provided (A_s)	Effective Area of Concrete ($b \times d$)	% of Steel Provided (ρ) = $A_s/(b \times d)$
3.16 in^2	12 × 17.625 = 211.5 in^2	3.16 × 100/211.5 = 1.49%

  From Table 1.6:

  for (ρ) of 1.49%, the strain in steel = 0.0067

  Since this strain is greater than 0.005, the section is tension controlled as desired.

- Minimum Area of Steel

  The minimum area of steel is given by:

Minimum Area of Steel	
$[(3\sqrt{f'_c})/f_y] \times (b \times d)$	$= (3\sqrt{4{,}000} / 60{,}000) \times (12 \times 17.625)$ $= 0.67$ in^2
but not less than: $(200/f_y) \times (b \times d)$	$= (200 / 60{,}000) \times (12 \times 17.625)$ $= 0.71$ in^2

  The area of steel provided (3.16 in^2) is indeed greater than the minimum area (0.71 in^2).

- Minimum Clearances for Steel

  For (4) #8 bars, the minimum clear rebar spacing shall be the greatest of:

  ✓ 1″
  ✓ one bar diameter = 1″
  ✓ 1.33 times the maximum aggregate size (assuming 3/4″ maximum aggregate size) 1.33 × 0.75″ = 0.997″

  } The minimum clear rebar spacing is therefore 1″

  Verify that steel reinforcing properly fits within assumed section:

  As shown in Figure 11.15, the minimum beam width is 11″, therefore our 12″ width assumption is adequate:

Clear cover	= 1.5″
Stirrup diameter	= 0.375″
Rebar size	= 1.0″
Min. clear rebar spacing	= 1.0″

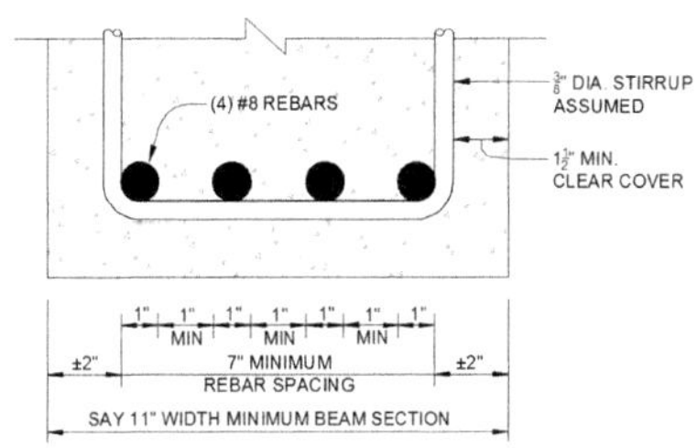

Figure 11.15 Minimum Width for Girder B

From Design Aid 1–6 in Appendix 3, we also see that for the given assumptions, a 12″-wide beam could accommodate four #8 rebars.

Step 2: Designing for Shear (Girder B)

Determine Shear Strength of Concrete Section (V_c)

$$\mathbf{V_c = 2\sqrt{f'_c} \times b \times d}$$
$$= [(2\sqrt{4,000}) \times 12 \times 17.625]/1,000$$
$$= 26.75\ k$$

(changing lbs to kips)

where:
- V_c = nominal shear strength of concrete
- f'_c = concrete compressive strength
- b = width of girder (12″)
- d = effective depth (17.625″)

Determine Shear Force to be Resisted by Steel (V_s)

$$\mathbf{V_s = (V_u / 0.75) - V_c}$$

where:
- V_s = nominal shear force to be resisted by steel
- V_u = maximum shear force = 26.0 k
- V_u @d = shear at distance (d) from face of support
- w_u = uniform load = 0.3 k/ft
- L = length of girder (16 ft)

The shear at a distance (d) from the face of the support will have a small reduction due only to the self-weight of the girder. It is determined by:

$$\mathbf{V_u@d = V_u - (w_u \times d / 12)}$$
$$= 26k - (0.3 \times 17.625/12)$$
$$= 25.56\ k$$

As we did for Beam 3, we'll take "d" to be the distance from the centerline of the support—a more conservative value (Figure 11.16).

Therefore:

$$V_s = (25.56/0.75) - 26.75$$
$$= 7.33\ k$$

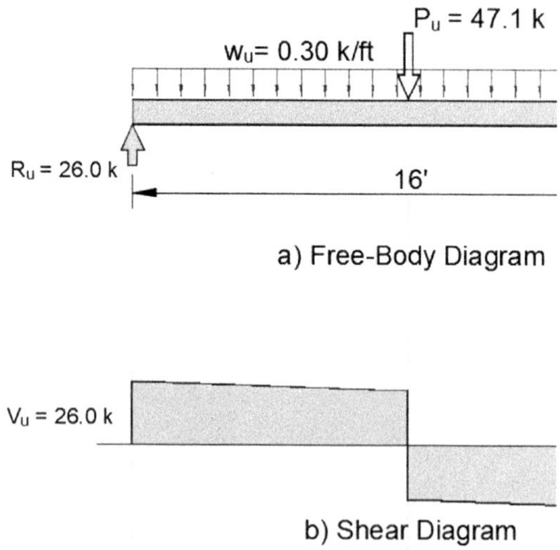

Figure 11.16 Shear at Distance 'd' From Support

Determine Stirrup Size and Spacing (s)

- Stirrups are required if:

$$
\begin{aligned}
V_u &> 0.5\ \Phi\ V_c & \text{where } V_u &= 26\text{ k}\\
&> 0.5 \times 0.75 \times 26.75 & V_c &= 26.75\text{ k}\\
&> 10.03\text{ k} & \Phi &= 0.75
\end{aligned}
$$

Since 26 k > 10.03 k, stirrups are required.

- Assume a stirrup rebar size (A_v)

Assume #3 with an area of 0.11 in^2 (Table 11.1).

- Calculate stirrup spacing (s)

$$\mathbf{s = (A_v \times f_y \times d) / V_s}$$

Since the stirrup is bent into a loop, the cross-sectional area of stirrup steel (A_v) is twice the cross-sectional area of the rebar, therefore: $\mathbf{A_v}$ = 2 × 0.11 = 0.22 in^2

$$
\begin{aligned}
s &= (0.22 \times 60 \times 17.625)/7.33\\
&= 206.25 / 7.33\\
&= 31.74''\ \text{o.c.}
\end{aligned}
$$

Check Limits for Reinforcement

- Verify that the shear capacity of steel is no greater than:

$$\mathbf{V_s < 4\sqrt{f'_c} \times b \times d}$$

$$(V_s) = 7.33\text{ k}$$

$$(4\sqrt{f'_c} \times b \times d) = [(4 \times \sqrt{4000}) \times 12 \times 17.625] / 1000 = 53.51\text{ k}$$

Since 7.33 k < 53.51 k, the section complies.

- Maximum stirrup spacing shall be the smallest of:

$$
\begin{aligned}
s_{max} &= (A_v \times f_y)/(50 \times b) = (0.22 \times 60{,}000)/(50 \times 12) = 22''\ \text{o.c.}\\
s_{max} &= d/2 = 17.625/2 = 8.8''\ \text{o.c.}\\
s_{max} &= 24''\ \text{o.c.}
\end{aligned}
$$

The smallest from the above is 8.8″ o.c.

We'll assume #3 stirrups @8″ o.c. along the entire length of the beam, with two #4 top rebar to hold the stirrups in place.

- Verify minimum area of stirrup:

Minimum Area of Stirrup		
$A_{vmin} = = 0.75\sqrt{f'_c}\,\frac{b \times s}{f_y}$	(0.75 √4000 × 12 × 8) / 60,000 =	0.085 in²
but not less than: $50\,\frac{b \times s}{f_y}$	(50 × 12 × 8) / 60,000 =	0.09 in²

The area of stirrup provided (A_v = 0.22 in²) exceeds the above minimums.

Step 3: Checking for Deflection (Girder B)

Since we've used span-to-depth ratios greater than the code-recommendations, deflections are considered to be kept within acceptable limits and need not be checked.

Figure 11.17 shows the final design for Girder B.

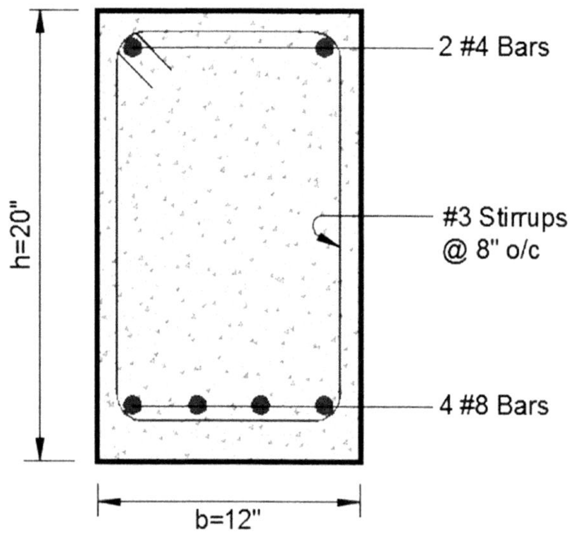

Figure 11.17 Girder B Final Design

A Word on Rebar Length

The Reader should note that the above exercise was for the purpose of calculating the quantity of rebar required for the section at maximum moment. Since the moment diminishes towards the supports, the quantity of steel required would also reduce. The ACI provides several guidelines for the minimum number of rebars that must continue, and how and where the others rebars can be terminated. This becomes a detailing exercise, not covered in this text.

11.4 COLUMN B2

General Assumptions

- f'_c = concrete compressive strength (4,000 psi)
- column is braced against sidesway
- f_y = yield strength of steel = 60,000 psi
- floor-to-floor column height = 12 ft

Determine Loads

First Floor Loads

Column B2 is supporting first floor Girders B and E and Beams 4 and 5. The combined load from the end reaction of each of these members is tabulated below and shown in Figure 11.18.

Member	First Floor Load on Column B2
Girder B	26.0 k
Girder E	26.0 k
Beam 5	23.5 k
Beam 6	23.5 k
Total	99.0 k

Roof Loads

In addition to the first floor loads, Column B2 is also supporting roof loads. Since the roof framing and loading is symmetrical about Column B2, we can calculate its roof load based on its load tributary area (Figure 11.19).

Load tributary area for Column B2 $= 16\text{ ft} \times 24\text{ ft} = 384\text{ sf}$

D = uniform roof dead load = 75 psf

L_r = uniform roof live load = 30 psf

Applying the controlling LRFD roof load combination:

$$1.2\,D + 1.6\,L_r = (1.2 \times 75) + (1.6 \times 30) = 138\text{ psf}$$

(see Chapter 4)

$$\text{Roof load on Column B2} = 384\text{ ft}^2 \times 138\text{ lb/ft}^2 = 52{,}992\text{ lb} = 53.0\text{ k}$$

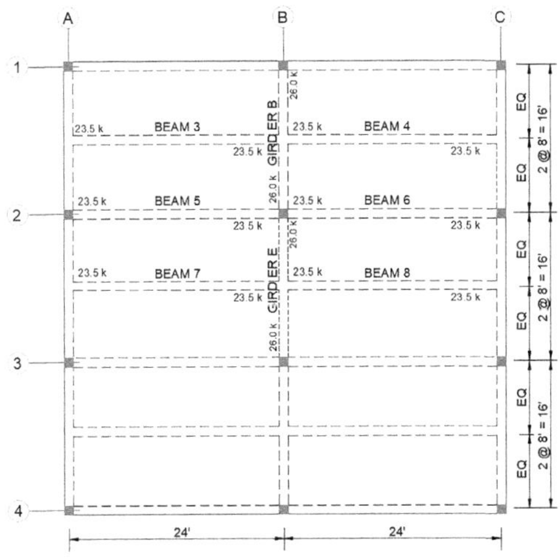

Figure 11.18 First Floor Loads on Column B2

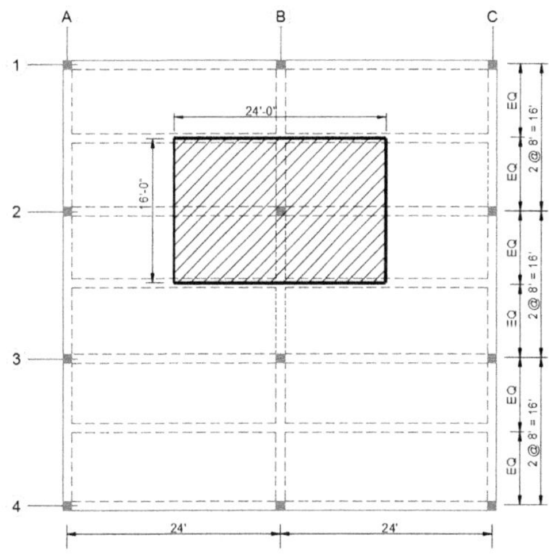

Figure 11.19 Roof Tributary Area for Column B2

Total Floor and Roof Loads on Column B2

First floor load	$= 99.0$ k
Roof load	$= 53.0$ k
Total Load on Column B2	$= 152.0$ k (required compressive strength, P_u)

Assume Concrete Cross-Sectional Shape and Dimensions

Column sizing is generally based on practical considerations such as minimizing size, simplifying formwork, and ease of rebar and concrete placement.

Let's assume a 12″ × 12″ column cross section consistent with the widths of Beam 3 and Girder B (Figure 11.20).

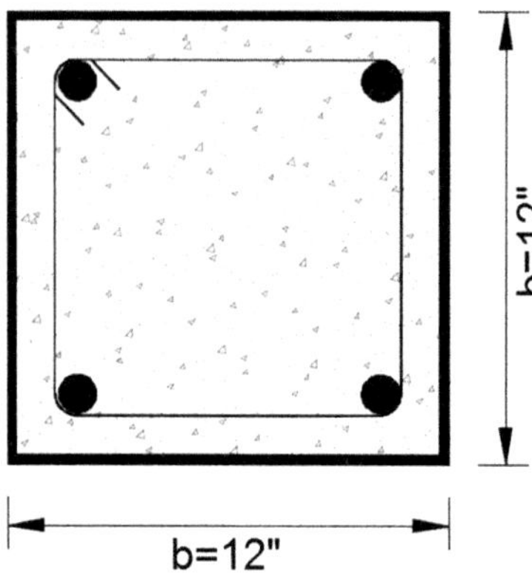

Figure 11.20 Column B2 Assumptions

Assume Vertical Steel

The ACI limits for the ratio of vertical reinforcing steel (A_{st}) to area of concrete (A_g) are:

- minimum percentage of steel: 1% of A_g
- maximum percentage of steel: 8% of A_g

Let's investigate the column capacity with the minimum requirement of 1%.

$$A_{st} = 1\% \text{ of } A_g = .01 \times 12 \times 12 = 1.44 \text{ in}^2$$

Using (4) rebars (code minimum for rectangular columns), the minimum area of each rebar is:

$$1.44 / 4 = 0.36 \text{ in}^2$$

The closest (but still larger) size of rebar is #6 with an area of 0.44 in^2 (see Table 11.1). Let's assume (4) #6 rebars with:

$$A_g = 4 \times 0.44 = 1.76 \text{ in}^2$$

Determine Lateral Ties

Since the vertical rebar are smaller than #10, we'll use #3 rebars (with an area of 0.11 in^2) for lateral ties.

Tie spacing shall be the smallest of:

48 × tie diameter =	48 × 3/8 =	18″
16 × size of main vertical rebar =	16 × 6/8 =	12″
least dimension of column =		12″

Since the smallest from the above is 12″, we'll use #3 lateral ties @12″ o.c. along the entire length of the column.

Check Slenderness Ratio

$$\begin{aligned}\text{Slenderness ratio} &= kl_u/r \\ &= 1\times(10.33\times12)/3.6 \\ &= 34.44\end{aligned}$$

Since $kl_u/r < 40$, slenderness effects need not be considered (i.e., the column may be considered short).

where:
'l' is the floor-to-floor height
'l_u' is the clear column height to underside of beam (i.e., the unbraced length)
Floor-to-floor height = 12 ft
Girder depth (20″) = 1.67 ft
Clear column height (l_u) = 12–1.67 = 10.33 ft
r = radius of gyration = 0.3 × 12 = 3.6″
(for rectangular columns, the ACI allows 'r' to be calculated as 0.3 × least dimension)

Determine Available Compressive Strength

The available compressive strength of a short column with tie reinforcement is given by:

$$\begin{aligned}\Phi P_n &= \Phi\times0.80\left[0.85 f'_c\left(A_g - A_{st}\right)+\left(A_{st}\times f_y\right)\right] \\ &= 0.65\times0.8\left[0.85\times4\ (144-1.76)+(1.76\times60)\right] \\ &= 0.65\times0.8\left[483.6+105.6\right] \\ &= 306.4\text{ k}\end{aligned}$$

where:
ΦP_n = available compressive strength
Φ = strength reduction factor for tied columns = 0.65
A_g = gross area of column cross section = 144 in^2
A_{st} = area of steel = 1.76 in^2

Verify Available Compressive Strength ≥ Required Compressive Strength

$$\Phi \; P_n \geq P_u$$

Available Compressive Strength (Φ P_n)	Required Compressive Strength (P_u)
306.4 k	152.0 k

Since Column B2's available compressive strength is significantly greater than that required, our column assumptions, with even the minimum percentage of steel, provide more than adequate capacity. Although we could reduce column dimensions, it's generally considered good practice to keep column, beam, and girder widths the same.

12 In Closing

In today's world, structural design is performed by computers using sophisticated software capable of analyzing hundreds of design variables in a short time. Given this capability, the student may reasonably ask why he/she should have to study the theory, concepts, and manual detail underlying the computer programs. Perhaps the best of many reasons is to equip the professional to be fluent in the language of structures—to competently evaluate concepts and assess solutions, whether performed by hand or by computer.

While design methodologies and approaches may evolve and change over time, structural principles remain constant. Our hope is that the previous chapters have given the Reader an appreciation of the basic properties and behavior of steel, wood, and reinforced concrete—the most commonly used materials in the construction industry. Our goal was to show the current design approaches for each of these materials, while relating them to their underlying principles. Our overall intent was to enrich the dialog between architect and engineer. If we have begun to accomplish these aims, our efforts will have been rewarded.

Glossary

The intent of this Glossary is to provide the Reader with a brief, concise description of terms. The Reader should refer to this text and/or other sources for a more complete technical explanation of terms.

A

Actual stress The stress in a structural member created by the design loads.

Adjusted design value The mathematical product of a 'reference design value' times the applicable 'adjustment factors'. Can be thought of as an 'allowable stress'.

Adjustment factors Account for variations in 'conditions of use' that a wood structure will encounter when erected. Adjustment factors are applied to a 'reference design value', resulting in an 'adjusted design value'.

Aggregate A concrete ingredient consisting of inert, clean, mineral matter such as sand, gravel, and crushed stone. Aggregate constitutes approximately 75% of the volume of concrete and can vary according to the type of mineral material available in any particular locale.

Allowable Strength Design (ASD) A design methodology used for steel, that assures a member's available strength (termed 'allowable strength' in ASD) is equal to or greater than the required strength.

Allowable strength The ASD term for 'available strength'.

Allowable Stress Design A design methodology used for wood, that assures a member's actual stress is less than or equal to the allowable stress.

Allowable stress The stress permitted in a structural member, after the application of safety factors.

Anisotropic A property of wood indicating that its structural characteristics vary depending upon the orientation of its grain.

Annular growth rings The outward seasonal growth of rings in a tree trunk that are produced by the manufacture of cells.

Available strength The strength of a structural member, after the application of safety factors.

Axial load A load evenly distributed about a column's longitudinal axis, tending to create pure compressive stress.

B

Balanced design In a reinforced concrete section, the point at which steel and concrete reach their permissible strains at the same time.

Beam A generally horizontal member that primarily resists bending.

Bending The tendency of a body, such as a beam, to bow.

Braced frame A structural frame made rigid by diagonal and/or cross bracing members at determined locations.

Brittle The material quality that results in sudden rupture with increasing stress, thereby giving no 'notice' before failure. Concrete and cast iron are two examples of materials having this quality.

Buckling A type of failure in a member caused by excessive compressive stress, and characterized by excessive bending.

C

Cambium A thin layer of living cells close to the outer surface of a tree trunk that manufactures wood cells.

Cantilever A section of a structural member, such as a beam, extending beyond its point of support.

Capacity A term synonymous with 'available strength'.

Center of gravity The point at which the mass of a three-dimensional body can be considered to be concentrated.

Centroid The point at which the area of a two-dimensional body can be considered to be concentrated.

Cold formed structural steel sections Produced by a series of shaping operations at relatively low temperatures that bend sheets of steel of specified thickness into the desired final configuration. Hollow structural sections (HSS), circular, square, and rectangular are common examples.

Column A generally vertical member that primarily resists compression, carrying loads downward to the foundations and ground. 'Pillar' and 'post' are two other terms for columns, usually implying shorter height.

Compact steel section A steel section with sufficient ratios for flange 'width-to-thickness' and web 'depth-to-thickness', to enable the section to achieve its full flexural strength before experiencing localized buckling failure.

Compression The tendency of a body to be crushed.

Concrete An artificial stone-like material made by mixing three basic ingredients—Portland cement, mineral aggregate, and water. Unreinforced concrete resists compressive stress well, but is virtually useless resisting tensile stress.

Concrete admixtures Various types may be added to concrete to modify its properties (such as increasing workability, reducing setting time, increasing setting time, etc.)

Concrete cover The thickness of concrete protecting steel reinforcing from fire and corrosion, and ensuring that the reinforcing is sufficiently embedded in the member to bond with the concrete.

Control joints Commonly placed in concrete slabs to induce cracking at pre-determined locations.

Creep The tendency of a material to continue to deform over time (i.e., exhibit strain), when placed under load for an extended period. Concrete is an example of a material that exhibits creep, albeit very slowly.

Critical load The load at which a column will fail, either by crushing or buckling.

Crushing A type of failure in a member caused by excessive compressive stress.

Curing The process of the water in a concrete mix evaporating as the mixture sets. Curing must be carefully controlled within specified moisture and temperature limits to avoid potential problems.

D

Dead load A load of fixed, constant, and predictable magnitude.

Deflection The maximum displacement (i.e., distance) from 'true' of a member experiencing bending.

Deformation The distorted shape of a member when stressed under load. Typically represented by a shortening, elongation, or bending of the member.

Design loads The loads obtained from the governing load combinations.

Design methodology A recognized approach to the design of structural members.

Design strength The LRFD and Strength Design term for 'available strength'.

Diaphragm Provides lateral resistance in a horizontal plane. If designed as such, building floors and roofs can act as diaphragms transferring horizontal lateral forces to vertical lateral resistance components, and ultimately to the ground.

Dimensional lumber Wood members nominally 4 inches or less in the least dimension.

Dressed lumber Wood that is seasoned and surfaced, resulting in member sizes slightly less than the nominal.

Ductile The material quality that results in plastic deformation when stressed beyond the yield point. Ductile materials 'stretch', thereby giving 'notice' before rupture. Steel is an example of a material having this quality.

Dynamic load A load considered as variable, such as an elevator or a vehicle in motion.

Dynamics The study of objects in motion.

E

Eccentric load A load unevenly distributed about a column's longitudinal axis, tending to create bending stresses in addition to compressive stresses.

Eccentricity The distance between the line of action of a vertical load, and the longitudinal axis of a column.

Effective length The theoretically adjusted length of a column as affected by its end restraints. A column's effective length is determined by its 'k-factor' which, for design purposes, can designate the column as greater than, less than, or equal to its actual length.

Elastic design Design theory based on the concept that resistance to bending is limited by the yielding of the extreme fibers in a member's cross section.

Elastic limit The point at which a material no longer returns to its original position when stress is removed, becoming permanently deformed,

Elastic range behavior The characteristics exhibited by a material wherein an increase in stress produces a proportional increase in strain (deformation), up to the material's 'elastic limit'. Within this range, strain disappears when stress is removed.

Elastic section modulus (S) Used in conjunction with elastic design and based on the concept that resistance to bending is limited by the yielding of the extreme fibers in a member's cross section.

End restraints (of a column) The type of support (idealized as pinned, fixed, or free) occurring at a column's top and bottom.

Engineered lumber Wood products manufactured by various industrial processes from wood pieces and components, to produce larger composite materials with consistent properties.

Environmental load A live load resulting from the effects of natural phenomena such as wind, earthquake, snow, etc.

Euler's formula Derived by the eighteenth century mathematician Leonhard Euler, it underlies column behavior theory. The formula defines the maximum vertical axial load that an 'ideal' column (i.e., one perfectly vertical, of constant cross section, and of homogeneous material) can support without buckling. The formula relates load, and column length, cross section and material.

Expansion joints Typically placed at determined intervals in structures and concrete slabs to allow a degree of movement.

F

Factored load A service load after the application of a load factor.

Fatigue The tendency of a material to weaken or fail, at a stress below the yield point, when the stress is repeatedly applied and removed. A paper clip that breaks after being bent back and forth several times is an example.

Fixity of connections Connections between concrete members, such as beams and columns, are typically cast monolithically with continuous reinforcing steel extending from one member to the other, thus developing a degree of 'fixity' (i.e., rigidity) and having the ability to resist moment.

Flexural cracks Cracks generally tending to occur in regions of maximum moment in concrete members.

Flexure A term used for bending.

Fly ash The by-product of burning coal in power generating plants.

Force As defined by Sir Isaac Newton in the 17th century, a force can cause, change, or stop motion on a body. A push, a pull, gravity, and friction are a few common examples of forces.

Frames Structures generally consisting of vertical and horizontal structural members, such as beams, girders, and columns.

Free-Body diagram A simplified representation of a structural member's loading and support conditions. It essentially shows the member isolated and 'cut free' from other members or conditions, which are then replaced by forces.

G

Girder A beam that supports other beams or members.

Glued laminated timber (Glulam) An engineered wood product made by bonding together individual layers of lumber that can be formed into any size or shape, including curves.

Grain orientation The alignment of the longitudinal cells in wood resulting in a 'direction of grain'. The direction of grain in a wood member is generally considered parallel to its length.

Gravity load A load that acts vertically towards the center of the earth.

H

Heartwood The portion of the tree trunk, outside the pith, containing dead cells that no longer serve a purpose except to support the tree.

Heavy timber construction An established construction method that uses heavy sawn or engineered wood timbers as structural members.

Hooke's Law Discovered by Robert Hooke in the late 1600s, it expresses the proportional relationship between stress and strain, up to a material's 'proportional limit'.

Hot rolled structural steel sections Produced by passing steel billets at very high temperatures through a succession of rollers that progressively squeeze the billets into the desired final configuration. Wide flange (W), channel (C), and angle (L) are common examples.

Hydration The chemical process of cement reacting with water in a concrete mix. Hydration is an exothermic process, meaning that it gives off heat which must be controlled during curing to avoid potential problems.

I

IBC (International Building Code) A building code established in 1994, that unified and replaced various other national codes in use at the time, Based on the IBC, cities and states develop and adopt their own governing codes that reflect their specific conditions and requirements.

In situ concrete Concrete produced on-site. Also referred to as 'site cast' and 'cast-in-place'.

Inflection point The point at which a member's bending changes from concave to convex, indicating a point of zero moment.

Iron The predominant ingredient in steel that is a basic element, naturally occurring in iron-rich rock formations called iron ore.

Iron pigs Cast iron bricks produced by the blast furnace process, ready for further processing into steel.

J

Joist A term for one of a series of beams, closely-spaced and of lighter weight. A steel joist typically refers to a lightweight truss member.

K

k-factor A constant applied to the actual length of a column, based on the column's end restraints. Applying a k-factor results in a theoretical 'effective length' that is used in the column's design.

klf kips per linear foot (k/ft)

ksf kips per square foot (k/ft^2)

ksi kips per square inch (k/in^2)

L

Laminated strand lumber (LSL) (see structural composite lumber - SCL)

Laminated veneer lumber (LVL) (see structural composite lumber - SCL)

Lateral load A load that acts horizontally; commonly wind and seismic loads.

Lateral resistance system The method of providing lateral rigidity in a frame structure, preventing its 'racking'. Braced frames, moment frames, and shear walls (or a combination thereof) are common methods used to provide lateral resistance for frame structures.

Least radius of gyration Determines the direction (i.e., axis) about which a column will tend to buckle. A column will tend to buckle about its axis having the least radius of gyration.

Limit-state design principles Define the boundaries of structural usefulness on which design methodologies are based, in terms of 'strength' and 'serviceability'.

Line of action (of a force) The infinite imaginary line passing through a vector force.

Lintel A type of beam generally placed over door and window openings.

Live load A load that is impermanent and unpredictable, and therefore given by codes.

Load and Resistance Factor Design (LRFD) A design methodology used for steel and wood that assures a member's available strength (termed 'design strength' in LRFD) is equal to or greater than the required strength.

Load combinations Account for the probability of various loads (such as dead, live, snow, rain, ice, wind, seismic, etc.) occurring simultaneously and/or at full magnitude.

Load factor Obtained from code-specified load combinations, it is applied to a service load essentially acting as a safety factor reflecting the uncertainty involved with it.

Load path The flow of loads through a structure to the ground.

Loads Forces imposed on a structure resulting from its use or from natural phenomena. Loads are broadly classified as dead, live, and environmental.

Lumber General term for wood utilized for structural purposes.

M

Mass timber construction A relatively new construction method gaining popularity and acceptance that uses thick, solid, large prefabricated wood panels as structural elements such as floors and walls. Two common types of mass timber panels are cross laminated timber (CLT), and nail laminated timber (NLT).

Member A general term typically used to designate a beam, girder, column, etc.

Modulus of elasticity A measure of a material's stiffness typically measured in psi or ksi. The higher the modulus of elasticity, the stiffer the material. Also known as Young's Modulus.

Moment The tendency of a force to produce rotational movement about a given point of rotation. Given by the magnitude of a force times its perpendicular distance to the point of rotation A measure of internal bending stress within a structural member, typically given in lb-feet or kip-feet.

Moment connection A method of attaching structural members that restricts rotational movement at the connection, thereby making it rigid.

Moment diagram A graphic depiction of the amount of internal bending stress (i.e., moment) at every section along a member's (typically a beam's) length.

Moment frame A structural frame made rigid by 'moment connections' between horizontal and vertical members at determined locations.

Moment of inertia (I) A cross-sectional property, from which other cross-sectional properties are derived, that is a measure of a cross section's stiffness. It is commonly used in deflection formulae and is typically given in inches^4.

N

Necking A phenomenon exhibited by ductile materials such as steel, characterized by localized stretching with a corresponding reduction in the cross sectional area. Once begun, with no increase in stress necking continues until rupture.

Neutral axis The imaginary theoretical line along the length of a member that exhibits neither compression nor tension when subjected to bending stress. In a member such as a beam, having a rectangular cross section of homogeneous material, the neutral axis occurs midway between the upper and lower edges.

Nominal Loosely defined as 'approximate'.

Nominal strength The strength of a structural member, before the application of safety factors.

P

Parallel strand lumber (PSL) see structural composite lumber (SCL)

Permanent set The permanent deformation that remains in a material when stressed beyond the elastic limit, even when the stress is removed.

Pig iron (see iron pigs)

Pin connection A method of attaching structural members that allows a degree of rotational movement at the connection, thereby not providing strict rigidity. A 'shear connection' in steel is considered to be a pin connection.

Pith At the very center of a tree trunk.

Plastic design Design theory based on the concept that resistance to bending is limited by the yielding of all the fibers in a material's cross section.

Plastic hinge The theoretical type of deformation that occurs at the point of maximum moment, when a member yields upon reaching its plastic limit.

Plastic limit The point at which all fibers in a cross section become fully and equally stressed to the yield point. In other words the member has reached its limit of material resistance.

Plastic moment The maximum moment that a member such as a beam can resist at its plastic limit.

Plastic range behavior The characteristics exhibited by a ductile material such as steel, when stressed beyond the elastic limit. Within this range, with relatively little increase in stress, strain increases disproportionately and the material behaves somewhat taffy-like. When stress is removed, the material no longer returns to its original position and becomes permanently deformed, the deformation being termed 'permanent set'.

Plastic section modulus (Z) Used in conjunction with plastic design and based on the concept that resistance to bending is limited by the yielding of all the fibers in a material's cross section.

plf pounds per linear foot (lb/ft)

Portland cement The ingredient in concrete that, when mixed with and activated by water, serves to bind concrete's materials into a rock-hard mass.

Pozzolans Also termed supplementary cementitious materials (SCMs), commonly include fly ash, silica fume, and slag - all waste by-products of various industrial processes. As cement substitutes, pozzolans are used to obtain higher strengths and reduce the quantity of cement in a mix.

Pre-cast concrete Concrete produced off-site in sophisticated plants and shipped to the construction site.

Prestressed concrete Concrete with induced initial compressive stresses, serving to counteract tensile stresses when placed under load. Pre-tensioning and post-tensioning are two methods of prestressing concrete.

Proportional limit The point beyond which the proportional relationship between stress and strain is no longer maintained in a material.

psf pounds per square foot (lb/ft^2)

psi pounds per square inch (lb/in^2)

Purlin A lightweight beam placed perpendicular to a roof rafter.

R

Racking The lateral distortion of a frame.

Radius of gyration (r) A cross-sectional property, derived from and proportional to moment of inertia, that is also a measure of a cross section's stiffness. It is commonly used in column design and typically given in inches.

Rafter An inclined roof beam.

Rebar Commonly-used term for steel reinforcing bars.

Reference design value A baseline stress for various wood grain orientations, based on species and grading, that takes safety factors into account.

Reinforced concrete Concrete have embedded steel reinforcing, primarily to resist tensile forces.

Required strength The strength needed by a structural member to support the design loads.

Resistance factor A factor applied to the nominal strength of a material in Load and Resistance Factor Design (LRFD), thereby reducing it and essentially serving as a safety factor.

S

Safety factors Provide a degree of reliability against failure by allowing for a 'margin of error'. Safety factors can be applied to a material's design strength (thereby reducing it) and/or to design loads (thereby generally increasing them),

Sapwood The portion of the tree trunk outside the heartwood that carries water and nutrients between roots and leaves.

Sawn lumber Wood cut and used directly from felled trees (logs),

Section When used in reference to a structural member, it implies its cross sectional shape.

Section modulus (S and Z) A cross-sectional property, derived from and proportional to moment of inertia, that is also a measure of a cross section's stiffness. There are two kinds of section moduli, elastic (S) and plastic (Z), commonly used in beam design and typically given in inches^3.

Seismic A term relating to earthquakes.

Service load A load estimation, without the application of a load factor.

Serviceability A measure of the 'comfort level' of building inhabitants involving among other things, excessive deflection and vibrations.

Shape factor An indication of the comparative moment capacity of a given cross section, in plastic vs. elastic design.

Shape When used in reference to a structural member, it is another term for 'section'.

Shear connection A term for a theoretical pin connection.

Shear cracks Cracks generally tending to occur in regions of maximum shear in concrete members.

Shear diagram A graphic depiction of the amount of internal shear stress at every section along a member's (typically a beam's) length.

Shear force The measurement of internal shear in a member, typically given in lbs or kips.

Shear walls Typically reinforced concrete or masonry walls designed to resist lateral shear forces as well as vertical gravity forces.

Shear The tendency of a body to be sliced. Shear in a member can be both perpendicular and parallel to its length.

Silica-fume The by-product of the manufacture of silicon. Also known as micro silica.

Slag The by-product of producing iron in blast furnaces.

Slenderness ratio A measure of a column's proportions as defined by its effective length divided by its least radius of gyration.

Static equilibrium A condition in which all forces and moments acting on a member or structure are 'in balance', keeping the member or structure from moving.

Static load A load analyzed as being constant and not moving.

Statics The study of objects at rest.

Steel An extremely strong alloy composed primarily of iron (mostly) and carbon.

Steel billets Steel bars processed from iron pigs, having the desired metallurgical and structural properties, ready for final processing into structural steel shapes or other uses.

Stiffness A measure of resistance to deformation (bending for example), commonly represented by modulus of elasticity for material stiffness, and moment of inertia for cross-sectional stiffness.

Stirrups The term for concrete shear reinforcing, typically steel rebar bent into vertically-placed loops.

Strain hardening range behavior The characteristics exhibited by a ductile material such as steel, when stressed beyond the plastic range. Within this range the material stabilizes somewhat and is able to take on additional stress with a corresponding increase in deformation (strain), until it reaches its ultimate strength and ruptures.

Strain The deformation in a member resulting from load, as measured against its original size. Strain has no unit of measurement.

Strength A structure or member's load carrying capability.

Strength Design A design methodology used for concrete, that assures a member's available strength (termed 'design strength' in Strength Design) is equal to or greater than the required strength.

Strength reduction factor A factor applied to the nominal strength of a material in Strength Design, thereby reducing it and essentially serving as a safety factor.

Strength-based design methodologies Involve assuring that a member's available strength is equal to or greater than the required strength.

Stress The internal force distribution on a member resulting from loads, typically expressed in psi or ksi.

Stress distribution diagram Shows the relative tensile and compressive stresses in a member's cross section, in relation to its neutral axis, at a particular location along the length of that member.

Stress-based design methodologies Involve assuring that a member's allowable stress is equal to or greater than the actual stress.

Stresses in wood Since grain orientation affects the structural properties of sawn wood, several types of stresses must be considered when designing in sawn wood.

Stress-Strain curve A graph plotting stress vs. strain that shows a material's behavioral characteristics under load. Every material has a unique stress-strain curve in tension as well as in compression, which can be considered its 'signature' behavior under load.

Structural composite lumber (SCL) Refers to a family of engineered wood products produced by bonding wood veneers or strands, into long blocks of material called billets. These are then cut to size in various sizes and thicknesses for use as structural members. Three common types of SCL are laminated veneer lumber (LVL) and laminated strand lumber (LSL) – used for lighter members, and parallel strand lumber (PSL – used for heavier members).

Structural design As applied to an individual structural member, the selection of an appropriate material and cross sectional shape to safely and economically resist the loads to which it will be subjected.

Structural engineering The branch of civil engineering concerned with designing buildings and other types of structures to stand up and resist loads,

Strut A short compression member, but not usually used in reference to a column.

T

Tension The tendency of a body to be pulled apart.

Tension controlled sections In the design of a member's concrete section, the goal is to have the reinforcing steel reach its maximum strain before the concrete, to avoid a sudden and catastrophic failure of the concrete. Such a concrete section is termed a tension controlled.

Timbers Wood members 5 inches or greater in the least dimension.

Torque A term similar in meaning to moment, typically used in mechanical engineering applications.

Torsion The tendency of a body to be twisted, resulting in shear stresses.

U

Ultimate strength The point at which a material ruptures (breaks).

V

Vector A force is an example of a vector in that it has magnitude (e.g., pounds, kips) and direction (e.g., up, down, left, right).

W

Web crippling The tendency of the web of a steel member to buckle under excessive shear forces.

Web stiffening The reinforcement of a steel member's web to prevent buckling.

Whitney stress distribution block A diagrammatic representation of the uniform compressive stresses exerted on the cross section of a concrete beam. Commonly referred to as the Whitney block.

Wood I-joists An engineered wood product typically consisting of a plywood or oriented strand board (OSB) web, attached to laminated veneer or sawn lumber flanges. Commonly used in floor construction in light wood frame construction and available in various depths and thicknesses.

Working Stress Design A design methodology term for Allowable Stress Design, used for certain types of concrete design.

Y

Yield point The point beyond which a deformed material will no longer return to its original unstressed condition upon removal of the stress. The material is considered to have essentially failed.

Yield stress The stress at which a material begins to exhibit excessive and permanent deformation, and can no longer perform as intended.

Young's modulus (see modulus of elasticity)

Appendix 1

The AISC Steel Construction Manual

The American Institute of Steel Construction (AISC) Manual, 15th Edition, is referenced in this text and contains guidelines for the design and use of steel. It consists of the following parts:

Part 1—Dimensions and Properties
Part 2—General Design Considerations
Part 3—Design of Flexural Members
Part 4—Design of Compression Members
Part 5—Design of Tension Members
Part 6—Design of Members Subject to Combined Loading
Part 7—Design Considerations for Bolts
Part 8—Design Considerations for Welds
Part 9—Design of Connecting Elements
Part 10—Design of Simple Shear Connections
Part 11—Design of Partially Restrained Moment Connections
Part 12—Design of Fully Restrained Moment Connections
Part 13—Design of Bracing Connections and Truss Connections
Part 14—Design of Beam Bearing Plates, Column Base Plates, Anchor Rods and Column Splices
Part 15—Design of Hanger Connections, Bracket Plates and Crane-rail Connections
Part 16—Specifications and Codes
Part 17—Miscellaneous Data and Mathematical Information

The parts of particular interest for the Reader are 1, 3, and 4. These parts have numerous tables and charts of structural shapes, organized in several ways to facilitate the selection of beams (flexural members) and columns (compression members) based on the user's selection criteria. For example, there are several different tables on steel shapes depending upon whether the designer is primarily interested in the shape's dimensions, the shape's design properties, selection by the shape's section modulus Z, or selection by the shape's ability to support load. For values pertaining to ASD, tabular columns are shown shaded. For values pertaining to LRFD, tabular columns are shown unshaded. Included here are excerpts of tables used in the Chapter 6 Case Study for Design in Steel. For a more complete explanation of these tables, the Reader is referred to the AISC Manual.

Table 1-1 W-Shapes—Dimensions, Properties

This table displays the dimensional and design properties of various W shapes. It contains information such as the shape's exact area, dimensions, moment of inertia, section modulus, radius of gyration, and torsional properties.

Table 3-2 W-Shapes—Selection by Z_x (F_y = 50 ksi)

This table facilitates the selection of a W beam based upon a desired section modulus, *Zx*. Among other information, this table displays a shape's available flexural strength (M_{px} / Ω_b and $\Phi_b M_{px}$) and available shear strength (V_{nx} / Ω_v and Φ_v V_{nx}) in ASD and LRFD respectively. Although not used in this text, the tabular columns labeled M_{rx} / Ω_b and Φ_b M_{rx} indicate a shape's available flexural strength in ASD and LRFD respectively, when the compression flange is not fully braced.

Table 3-6 Maximum Total Uniform Load, kips—W-Shapes (F_y = 50 ksi)

This table facilitates the selection of simply supported W beams carrying uniformly distributed loads, by listing the total load (wL) that the beam can carry. The user enters the table with the length (span) of the beam and selects a beam based on ASD or LRFD.

Table 4-1 Available Strength in Axial Compression, kips—W-Shapes (F_y = 50 ksi)

This table facilitates the selection of a W column based upon a particular kl/r ratio. It provides the axial capacity (i.e., available strength in axial compression) of a column, based upon buckling about the weaker yy axis, in ASD (P_n / Ω_c) and LRFD ($\Phi_c P_n$).

Table 4-4 Available Strength in Axial Compression, kips—Square HSS (F_y = 50 ksi)

This table facilitates the selection of a square HSS column based upon a particular kl/r ratio. It is analogous to Table 4-1 with the exception that both xx and yy axes have the same strength (i.e., there is no weaker axis).

Table 4-14 Available Critical Stress for Compression Members

This table displays the available critical stress of compression members (F_{cr} / Ω_c and $\Phi_c F_{cr}$ in ASD and LRFD respectively), based upon a particular kl/r ratio, for various common grades of steel. Consistent with Euler's formula, it shows the critical compressive stress decreasing as the column's kl/r ratio increases.

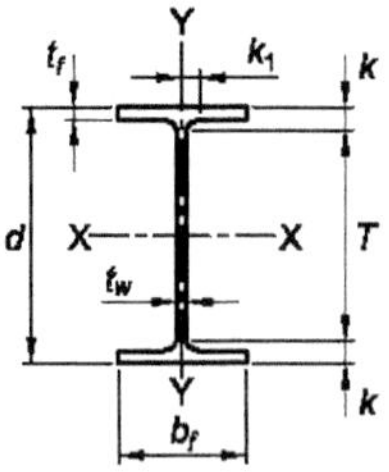

Table 1-1 (continued)
W-Shapes
Dimensions

Shape	Area, A	Depth, d		Web: Thickness, t_w		Web: $\frac{t_w}{2}$	Flange: Width, b_f		Flange: Thickness, t_f		Distance: k k_{des}	Distance: k k_{det}	k_1	T	Workable Gage
	in.^{2}	in.		in.		in.	in.		in.		in.	in.	in.	in.	in.
W14×38[c]	11.2	14.1	14 1/8	0.310	5/16	3/16	6.77	6 3/4	0.515	1/2	0.915	1 1/4	13/16	11 5/8	3 1/2[g]
×34[c]	10.0	14.0	14	0.285	5/16	3/16	6.75	6 3/4	0.455	7/16	0.855	1 3/16	3/4	↓	3 1/2
×30[c]	8.85	13.8	13 7/8	0.270	1/4	1/8	6.73	6 3/4	0.385	3/8	0.785	1 1/8	3/4	↓	3 1/2
W14×26[c]	7.69	13.9	13 7/8	0.255	1/4	1/8	5.03	5	0.420	7/16	0.820	1 1/8	3/4	11 5/8	2 3/4[g]
×22[c]	6.49	13.7	13 3/4	0.230	1/4	1/8	5.00	5	0.335	5/16	0.735	1 1/16	3/4	11 5/8	2 3/4[g]
W12×336[h]	98.9	16.8	16 7/8	1.78	1 3/4	7/8	13.4	13 3/8	2.96	2 15/16	3.55	3 7/8	1 11/16	9 1/8	5 1/2
×305[h]	89.5	16.3	16 3/8	1.63	1 5/8	13/16	13.2	13 1/4	2.71	2 11/16	3.30	3 5/8	1 5/8		
×279[h]	81.9	15.9	15 7/8	1.53	1 1/2	3/4	13.1	13 1/8	2.47	2 1/2	3.07	3 3/8	1 5/8		
×252[h]	74.1	15.4	15 3/8	1.40	1 3/8	11/16	13.0	13	2.25	2 1/4	2.85	3 1/8	1 1/2		
×230[h]	67.7	15.1	15	1.29	1 5/16	11/16	12.9	12 7/8	2.07	2 1/16	2.67	2 15/16	1 1/2		
×210	61.8	14.7	14 3/4	1.18	1 3/16	5/8	12.8	12 3/4	1.90	1 7/8	2.50	2 13/16	1 7/16		
×190	56.0	14.4	14 3/8	1.06	1 1/16	9/16	12.7	12 5/8	1.74	1 3/4	2.33	2 5/8	1 3/8		
×170	50.0	14.0	14	0.960	15/16	1/2	12.6	12 5/8	1.56	1 9/16	2.16	2 7/16	1 5/16		
×152	44.7	13.7	13 3/4	0.870	7/8	7/16	12.5	12 1/2	1.40	1 3/8	2.00	2 5/16	1 1/4		
×136	39.9	13.4	13 3/8	0.790	13/16	7/16	12.4	12 3/8	1.25	1 1/4	1.85	2 1/8	1 1/4		
×120	35.2	13.1	13 1/8	0.710	11/16	3/8	12.3	12 3/8	1.11	1 1/8	1.70	2	1 3/16		
×106	31.2	12.9	12 7/8	0.610	5/8	5/16	12.2	12 1/4	0.990	1	1.59	1 7/8	1 1/8		
×96	28.2	12.7	12 3/4	0.550	9/16	5/16	12.2	12 1/8	0.900	7/8	1.50	1 13/16	1 1/8		
×87	25.6	12.5	12 1/2	0.515	1/2	1/4	12.1	12 1/8	0.810	13/16	1.41	1 11/16	1 1/16		
×79	23.2	12.4	12 3/8	0.470	1/2	1/4	12.1	12 1/8	0.735	3/4	1.33	1 5/8	1 1/16		
×72	21.1	12.3	12 1/4	0.430	7/16	1/4	12.0	12	0.670	11/16	1.27	1 9/16	1 1/16		
×65[f]	19.1	12.1	12 1/8	0.390	3/8	3/16	12.0	12	0.605	5/8	1.20	1 1/2	1	↓	↓
W12×58	17.0	12.2	12 1/4	0.360	3/8	3/16	10.0	10	0.640	5/8	1.24	1 1/2	15/16	9 1/4	5 1/2
×53	15.6	12.1	12	0.345	3/8	3/16	10.0	10	0.575	9/16	1.18	1 3/8	15/16	9 1/4	5 1/2
W12×50	14.6	12.2	12 1/4	0.370	3/8	3/16	8.08	8 1/8	0.640	5/8	1.14	1 1/2	15/16	9 1/4	5 1/2
×45	13.1	12.1	12	0.335	5/16	3/16	8.05	8	0.575	9/16	1.08	1 3/8	15/16		
×40	11.7	11.9	12	0.295	5/16	3/16	8.01	8	0.515	1/2	1.02	1 3/8	7/8	↓	↓
W12×35[c]	10.3	12.5	12 1/2	0.300	5/16	3/16	6.56	6 1/2	0.520	1/2	0.820	1 3/16	3/4	10 1/8	3 1/2
×30[c]	8.79	12.3	12 3/8	0.260	1/4	1/8	6.52	6 1/2	0.440	7/16	0.740	1 1/8	3/4		
×26[c]	7.65	12.2	12 1/4	0.230	1/4	1/8	6.49	6 1/2	0.380	3/8	0.680	1 1/16	3/4	↓	↓
W12×22[c]	6.48	12.3	12 1/4	0.260	1/4	1/8	4.03	4	0.425	7/16	0.725	15/16	5/8	10 3/8	2 1/4[g]
×19[c]	5.57	12.2	12 1/8	0.235	1/4	1/8	4.01	4	0.350	3/8	0.650	7/8	9/16		
×16[c]	4.71	12.0	12	0.220	1/4	1/8	3.99	4	0.265	1/4	0.565	13/16	9/16		
×14[c,v]	4.16	11.9	11 7/8	0.200	3/16	1/8	3.97	4	0.225	1/4	0.525	3/4	9/16	↓	↓

[c] Shape is slender for compression with F_y = 50 ksi.
[f] Shape exceeds compact limit for flexure with F_y = 50 ksi.
[g] The actual size, combination and orientation of fastener components should be compared with the geometry of the cross section to ensure compatibility.
[h] Flange thickness greater than 2 in. Special requirements may apply per AISC *Specification* Section A3.1c.
[v] Shape does not meet the h/t_w limit for shear in AISC *Specification* Section G2.1(a) with F_y = 50 ksi.

Table 1-1 (continued)
W-Shapes
Properties

W14–W12

Nominal Wt.	Compact Section Criteria		Axis X-X				Axis Y-Y				r_{ts}	h_o	Torsional Properties	
	$\frac{b_f}{2t_f}$	$\frac{h}{t_w}$	I	S	r	Z	I	S	r	Z			J	C_w
lb/ft			in.^{4}	in.^{3}	in.	in.^{3}	in.^{4}	in.^{3}	in.	in.^{3}	in.	in.	in.^{4}	in.^{6}
38	6.57	39.6	385	54.6	5.87	61.5	26.7	7.88	1.55	12.1	1.82	13.6	0.798	1230
34	7.41	43.1	340	48.6	5.83	54.6	23.3	6.91	1.53	10.6	1.80	13.5	0.569	1070
30	8.74	45.4	291	42.0	5.73	47.3	19.6	5.82	1.49	8.99	1.77	13.4	0.380	887
26	5.98	48.1	245	35.3	5.65	40.2	8.91	3.55	1.08	5.54	1.30	13.5	0.358	405
22	7.46	53.3	199	29.0	5.54	33.2	7.00	2.80	1.04	4.39	1.27	13.4	0.208	314
336	2.26	5.47	4060	483	6.41	603	1190	177	3.47	274	4.13	13.8	243	57000
305	2.45	5.98	3550	435	6.29	537	1050	159	3.42	244	4.05	13.6	185	48600
279	2.66	6.35	3110	393	6.16	481	937	143	3.38	220	4.00	13.4	143	42000
252	2.89	6.96	2720	353	6.06	428	828	127	3.34	196	3.93	13.2	108	35800
230	3.11	7.56	2420	321	5.97	386	742	115	3.31	177	3.87	13.0	83.8	31200
210	3.37	8.23	2140	292	5.89	348	664	104	3.28	159	3.81	12.8	64.7	27200
190	3.65	9.16	1890	263	5.82	311	589	93.0	3.25	143	3.77	12.7	48.8	23600
170	4.03	10.1	1650	235	5.74	275	517	82.3	3.22	126	3.70	12.4	35.6	20100
152	4.46	11.2	1430	209	5.66	243	454	72.8	3.19	111	3.66	12.3	25.8	17200
136	4.96	12.3	1240	186	5.58	214	398	64.2	3.16	98.0	3.61	12.2	18.5	14700
120	5.57	13.7	1070	163	5.51	186	345	56.0	3.13	85.4	3.56	12.0	12.9	12400
106	6.17	15.9	933	145	5.47	164	301	49.3	3.11	75.1	3.52	11.9	9.13	10700
96	6.76	17.7	833	131	5.44	147	270	44.4	3.09	67.5	3.49	11.8	6.85	9410
87	7.48	18.9	740	118	5.38	132	241	39.7	3.07	60.4	3.46	11.7	5.10	8270
79	8.22	20.7	662	107	5.34	119	216	35.8	3.05	54.3	3.43	11.7	3.84	7330
72	8.99	22.6	597	97.4	5.31	108	195	32.4	3.04	49.2	3.41	11.6	2.93	6540
65	9.92	24.9	533	87.9	5.28	96.8	174	29.1	3.02	44.1	3.38	11.5	2.18	5780
58	7.82	27.0	475	78.0	5.28	86.4	107	21.4	2.51	32.5	2.81	11.6	2.10	3570
53	8.69	28.1	425	70.6	5.23	77.9	95.8	19.2	2.48	29.1	2.79	11.5	1.58	3160
50	6.31	26.8	391	64.2	5.18	71.9	56.3	13.9	1.96	21.3	2.25	11.6	1.71	1880
45	7.00	29.6	348	57.7	5.15	64.2	50.0	12.4	1.95	19.0	2.23	11.5	1.26	1650
40	7.77	33.6	307	51.5	5.13	57.0	44.1	11.0	1.94	16.8	2.21	11.4	0.906	1440
35	6.31	36.2	285	45.6	5.25	51.2	24.5	7.47	1.54	11.5	1.79	12.0	0.741	879
30	7.41	41.8	238	38.6	5.21	43.1	20.3	6.24	1.52	9.56	1.77	11.9	0.457	720
26	8.54	47.2	204	33.4	5.17	37.2	17.3	5.34	1.51	8.17	1.75	11.8	0.300	607
22	4.74	41.8	156	25.4	4.91	29.3	4.66	2.31	0.848	3.66	1.04	11.9	0.293	164
19	5.72	46.2	130	21.3	4.82	24.7	3.76	1.88	0.822	2.98	1.02	11.9	0.180	131
16	7.53	49.4	103	17.1	4.67	20.1	2.82	1.41	0.773	2.26	0.983	11.7	0.103	96.9
14	8.82	54.3	88.6	14.9	4.62	17.4	2.36	1.19	0.753	1.90	0.961	11.7	0.0704	80.4

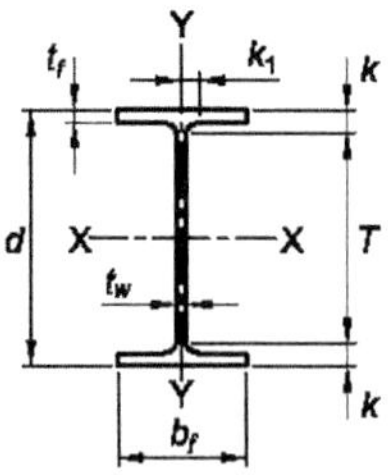

Table 1-1 (continued)
W-Shapes
Dimensions

Shape	Area, A	Depth, d		Web Thickness, t_w		Web $\frac{t_w}{2}$	Flange Width, b_f		Flange Thickness, t_f		Distance k k_{des}	Distance k k_{det}	k_1	T	Workable Gage
	in.^{2}	in.		in.		in.	in.		in.		in.	in.	in.	in.	in.
W10×112	32.9	11.4	11 3/8	0.755	3/4	3/8	10.4	10 3/8	1.25	1 1/4	1.75	1 15/16	1	7 1/2	5 1/2
×100	29.3	11.1	11 1/8	0.680	11/16	3/8	10.3	10 3/8	1.12	1 1/8	1.62	1 13/16	1	↓	↓
×88	26.0	10.8	10 7/8	0.605	5/8	5/16	10.3	10 1/4	0.990	1	1.49	1 11/16	15/16	↓	↓
×77	22.7	10.6	10 5/8	0.530	1/2	1/4	10.2	10 1/4	0.870	7/8	1.37	1 9/16	7/8	↓	↓
×68	19.9	10.4	10 3/8	0.470	1/2	1/4	10.1	10 1/8	0.770	3/4	1.27	1 7/16	7/8	↓	↓
×60	17.7	10.2	10 1/4	0.420	7/16	1/4	10.1	10 1/8	0.680	11/16	1.18	1 3/8	13/16	↓	↓
×54	15.8	10.1	10 1/8	0.370	3/8	3/16	10.0	10	0.615	5/8	1.12	1 5/16	13/16	↓	↓
×49	14.4	10.0	10	0.340	5/16	3/16	10.0	10	0.560	9/16	1.06	1 1/4	13/16	↓	↓
W10×45	13.3	10.1	10 1/8	0.350	3/8	3/16	8.02	8	0.620	5/8	1.12	1 5/16	13/16	7 1/2	5 1/2
×39	11.5	9.92	9 7/8	0.315	5/16	3/16	7.99	8	0.530	1/2	1.03	1 3/16	13/16	↓	↓
×33	9.71	9.73	9 3/4	0.290	5/16	3/16	7.96	8	0.435	7/16	0.935	1 1/8	3/4	↓	↓
W10×30	8.84	10.5	10 1/2	0.300	5/16	3/16	5.81	5 3/4	0.510	1/2	0.810	1 1/8	11/16	8 1/4	2 3/4[g]
×26	7.61	10.3	10 3/8	0.260	1/4	1/8	5.77	5 3/4	0.440	7/16	0.740	1 1/16	11/16	↓	↓
×22[c]	6.49	10.2	10 1/8	0.240	1/4	1/8	5.75	5 3/4	0.360	3/8	0.660	15/16	5/8	↓	↓
W10×19	5.62	10.2	10 1/4	0.250	1/4	1/8	4.02	4	0.395	3/8	0.695	15/16	5/8	8 3/8	2 1/4[g]
×17[c]	4.99	10.1	10 1/8	0.240	1/4	1/8	4.01	4	0.330	5/16	0.630	7/8	9/16	↓	↓
×15[c]	4.41	9.99	10	0.230	1/4	1/8	4.00	4	0.270	1/4	0.570	13/16	9/16	↓	↓
×12[c,f]	3.54	9.87	9 7/8	0.190	3/16	1/8	3.96	4	0.210	3/16	0.510	3/4	9/16	↓	↓
W8×67	19.7	9.00	9	0.570	9/16	5/16	8.28	8 1/4	0.935	15/16	1.33	1 5/8	15/16	5 3/4	5 1/2
×58	17.1	8.75	8 3/4	0.510	1/2	1/4	8.22	8 1/4	0.810	13/16	1.20	1 1/2	7/8	↓	↓
×48	14.1	8.50	8 1/2	0.400	3/8	3/16	8.11	8 1/8	0.685	11/16	1.08	1 3/8	13/16	↓	↓
×40	11.7	8.25	8 1/4	0.360	3/8	3/16	8.07	8 1/8	0.560	9/16	0.954	1 1/4	13/16	↓	↓
×35	10.3	8.12	8 1/8	0.310	5/16	3/16	8.02	8	0.495	1/2	0.889	1 3/16	13/16	↓	↓
×31[f]	9.13	8.00	8	0.285	5/16	3/16	8.00	8	0.435	7/16	0.829	1 1/8	3/4	↓	↓
W8×28	8.25	8.06	8	0.285	5/16	3/16	6.54	6 1/2	0.465	7/16	0.859	15/16	5/8	6 1/8	4
×24	7.08	7.93	7 7/8	0.245	1/4	1/8	6.50	6 1/2	0.400	3/8	0.794	7/8	9/16	6 1/8	4
W8×21	6.16	8.28	8 1/4	0.250	1/4	1/8	5.27	5 1/4	0.400	3/8	0.700	7/8	9/16	6 1/2	2 3/4[g]
×18	5.26	8.14	8 1/8	0.230	1/4	1/8	5.25	5 1/4	0.330	5/16	0.630	13/16	9/16	6 1/2	2 3/4[g]
W8×15	4.44	8.11	8 1/8	0.245	1/4	1/8	4.02	4	0.315	5/16	0.615	13/16	9/16	6 1/2	2 1/4[g]
×13	3.84	7.99	8	0.230	1/4	1/8	4.00	4	0.255	1/4	0.555	3/4	9/16	↓	↓
×10[c,f]	2.96	7.89	7 7/8	0.170	3/16	1/8	3.94	4	0.205	3/16	0.505	11/16	1/2	↓	↓

[c] Shape is slender for compression with $F_y = 50$ ksi.
[f] Shape exceeds compact limit for flexure with $F_y = 50$ ksi.
[g] The actual size, combination and orientation of fastener components should be compared with the geometry of the cross section to ensure compatibility.

Table 1-1 (continued)
W-Shapes
Properties

W10–W8

Nom-inal Wt.	Compact Section Criteria		Axis X-X				Axis Y-Y				r_{ts}	h_o	Torsional Properties	
	$\frac{b_f}{2t_f}$	$\frac{h}{t_w}$	I	S	r	Z	I	S	r	Z			J	C_w
lb/ft			in.^{4}	in.^{3}	in.	in.^{3}	in.^{4}	in.^{3}	in.	in.^{3}	in.	in.	in.^{4}	in.^{6}
112	4.17	10.4	716	126	4.66	147	236	45.3	2.68	69.2	3.08	10.2	15.1	6020
100	4.62	11.6	623	112	4.60	130	207	40.0	2.65	61.0	3.04	10.0	10.9	5150
88	5.18	13.0	534	98.5	4.54	113	179	34.8	2.63	53.1	2.99	9.81	7.53	4330
77	5.86	14.8	455	85.9	4.49	97.6	154	30.1	2.60	45.9	2.95	9.73	5.11	3630
68	6.58	16.7	394	75.7	4.44	85.3	134	26.4	2.59	40.1	2.92	9.63	3.56	3100
60	7.41	18.7	341	66.7	4.39	74.6	116	23.0	2.57	35.0	2.88	9.52	2.48	2640
54	8.15	21.2	303	60.0	4.37	66.6	103	20.6	2.56	31.3	2.85	9.49	1.82	2320
49	8.93	23.1	272	54.6	4.35	60.4	93.4	18.7	2.54	28.3	2.84	9.44	1.39	2070
45	6.47	22.5	248	49.1	4.32	54.9	53.4	13.3	2.01	20.3	2.27	9.48	1.51	1200
39	7.53	25.0	209	42.1	4.27	46.8	45.0	11.3	1.98	17.2	2.24	9.39	0.976	992
33	9.15	27.1	171	35.0	4.19	38.8	36.6	9.20	1.94	14.0	2.20	9.30	0.583	791
30	5.70	29.5	170	32.4	4.38	36.6	16.7	5.75	1.37	8.84	1.60	9.99	0.622	414
26	6.56	34.0	144	27.9	4.35	31.3	14.1	4.89	1.36	7.50	1.58	9.86	0.402	345
22	7.99	36.9	118	23.2	4.27	26.0	11.4	3.97	1.33	6.10	1.55	9.84	0.239	275
19	5.09	35.4	96.3	18.8	4.14	21.6	4.29	2.14	0.874	3.35	1.06	9.81	0.233	104
17	6.08	36.9	81.9	16.2	4.05	18.7	3.56	1.78	0.845	2.80	1.04	9.77	0.156	85.1
15	7.41	38.5	68.9	13.8	3.95	16.0	2.89	1.45	0.810	2.30	1.01	9.72	0.104	68.3
12	9.43	46.6	53.8	10.9	3.90	12.6	2.18	1.10	0.785	1.74	0.983	9.66	0.0547	50.9
67	4.43	11.1	272	60.4	3.72	70.1	88.6	21.4	2.12	32.7	2.43	8.07	5.05	1440
58	5.07	12.4	228	52.0	3.65	59.8	75.1	18.3	2.10	27.9	2.39	7.94	3.33	1180
48	5.92	15.9	184	43.2	3.61	49.0	60.9	15.0	2.08	22.9	2.35	7.82	1.96	931
40	7.21	17.6	146	35.5	3.53	39.8	49.1	12.2	2.04	18.5	2.31	7.69	1.12	726
35	8.10	20.5	127	31.2	3.51	34.7	42.6	10.6	2.03	16.1	2.28	7.63	0.769	619
31	9.19	22.3	110	27.5	3.47	30.4	37.1	9.27	2.02	14.1	2.26	7.57	0.536	530
28	7.03	22.3	98.0	24.3	3.45	27.2	21.7	6.63	1.62	10.1	1.84	7.60	0.537	312
24	8.12	25.9	82.7	20.9	3.42	23.1	18.3	5.63	1.61	8.57	1.81	7.53	0.346	259
21	6.59	27.5	75.3	18.2	3.49	20.4	9.77	3.71	1.26	5.69	1.46	7.88	0.282	152
18	7.95	29.9	61.9	15.2	3.43	17.0	7.97	3.04	1.23	4.66	1.43	7.81	0.172	122
15	6.37	28.1	48.0	11.8	3.29	13.6	3.41	1.70	0.876	2.67	1.06	7.80	0.137	51.8
13	7.84	29.9	39.6	9.91	3.21	11.4	2.73	1.37	0.843	2.15	1.03	7.74	0.0871	40.8
10	9.61	40.5	30.8	7.81	3.22	8.87	2.09	1.06	0.841	1.66	1.01	7.69	0.0426	30.9

Z_x

Table 3-2 (continued)
W-Shapes
Selection by Z_x

F_y = 50 ksi

Shape	Z_x	M_{px}/Ω_b	$\phi_b M_{px}$	M_{rx}/Ω_b	$\phi_b M_{rx}$	BF/Ω_b	$\phi_b BF$	L_p	L_r	I_x	V_{nx}/Ω_v	$\phi_v V_{nx}$
		kip-ft	kip-ft	kip-ft	kip-ft	kips	kips				kips	kips
	in.^{3}	ASD	LRFD	ASD	LRFD	ASD	LRFD	ft	ft	in.^{4}	ASD	LRFD
W21×44	**95.4**	**238**	358	**143**	214	**11.1**	16.8	**4.45**	**13.0**	**843**	**145**	217
W16×50	92.0	230	345	141	213	7.69	11.4	5.62	17.2	659	124	186
W18×46	90.7	226	340	138	207	9.63	14.6	4.56	13.7	712	130	195
W14×53	87.1	217	327	136	204	5.22	7.93	6.78	22.3	541	103	154
W12×58	86.4	216	324	136	205	3.82	5.69	8.87	29.8	475	87.8	132
W10×68	85.3	213	320	132	199	2.58	3.85	9.15	40.6	394	97.8	147
W16×45	82.3	205	309	127	191	7.12	10.8	5.55	16.5	586	111	167
W18×40	**78.4**	**196**	294	**119**	180	**8.94**	13.2	**4.49**	**13.1**	**612**	**113**	169
W14×48	78.4	196	294	123	184	5.09	7.67	6.75	21.1	484	93.8	141
W12×53	77.9	194	292	123	185	3.65	5.50	8.76	28.2	425	83.5	125
W10×60	74.6	186	280	116	175	2.54	3.82	9.08	36.6	341	85.7	129
W16×40	**73.0**	**182**	274	**113**	170	**6.67**	10.0	**5.55**	**15.9**	**518**	**97.6**	146
W12×50	71.9	179	270	112	169	3.97	5.98	6.92	23.8	391	90.3	135
W8×67	70.1	175	263	105	159	1.75	2.59	7.49	47.6	272	103	154
W14×43	69.6	174	261	109	164	4.88	7.28	6.68	20.0	428	83.6	125
W10×54	66.6	166	250	105	158	2.48	3.75	9.04	33.6	303	74.7	112
W18×35	**66.5**	**166**	249	**101**	151	**8.14**	12.3	**4.31**	**12.3**	**510**	**106**	159
W12×45	64.2	160	241	101	151	3.80	5.80	6.89	22.4	348	81.1	122
W16×36	64.0	160	240	98.7	148	6.24	9.36	5.37	15.2	448	93.8	141
W14×38	61.5	153	231	95.4	143	5.37	8.20	5.47	16.2	385	87.4	131
W10×49	60.4	151	227	95.4	143	2.46	3.71	8.97	31.6	272	68.0	102
W8×58	59.8	149	224	90.8	137	1.70	2.55	7.42	41.6	228	89.3	134
W12×40	57.0	142	214	89.9	135	3.66	5.54	6.85	21.1	307	70.2	105
W10×45	54.9	137	206	85.8	129	2.59	3.89	7.10	26.9	248	70.7	106
W14×34	**54.6**	**136**	205	**84.9**	128	**5.01**	7.55	**5.40**	**15.6**	**340**	**79.8**	120
W16×31	**54.0**	**135**	203	**82.4**	124	**6.86**	10.3	**4.13**	**11.8**	**375**	**87.5**	131
W12×35	51.2	128	192	79.6	120	4.34	6.45	5.44	16.6	285	75.0	113
W8×48	49.0	122	184	75.4	113	1.67	2.55	7.35	35.2	184	68.0	102
W14×30	**47.3**	**118**	177	**73.4**	110	**4.63**	6.95	**5.26**	**14.9**	**291**	**74.5**	112
W10×39	46.8	117	176	73.5	111	2.53	3.78	6.99	24.2	209	62.5	93.7
W16×26[v]	**44.2**	**110**	166	**67.1**	101	**5.93**	8.98	**3.96**	**11.2**	**301**	**70.5**	106
W12×30	43.1	108	162	67.4	101	3.97	5.96	5.37	15.6	238	64.0	95.9

ASD	LRFD
Ω_b = 1.67	ϕ_b = 0.90
Ω_v = 1.50	ϕ_v = 1.00

[v] Shape does not meet the h/t_w limit for shear in AISC *Specification* Section G2.1(a) with F_y = 50 ksi; therefore, ϕ_v = 0.90 and Ω_v = 1.67.

F_y = 50 ksi

Table 3-2 (continued)
W-Shapes
Selection by Z_x

Z_X

Shape	Z_x	M_{px}/Ω_b	$\phi_b M_{px}$	M_{rx}/Ω_b	$\phi_b M_{rx}$	BF/Ω_b	$\phi_b BF$	L_p	L_r	I_x	V_{nx}/Ω_v	$\phi_v V_{nx}$
		kip-ft	kip-ft	kip-ft	kip-ft	kips	kips				kips	kips
	in.^{3}	ASD	LRFD	ASD	LRFD	ASD	LRFD	ft	ft	in.^{4}	ASD	LRFD
W14×26	**40.2**	**100**	151	**61.7**	92.7	**5.33**	8.11	**3.81**	**11.0**	**245**	**70.9**	106
W8×40	39.8	99.3	149	62.0	93.2	1.64	2.46	7.21	29.9	146	59.4	89.1
W10×33	38.8	96.8	146	61.1	91.9	2.39	3.62	6.85	21.8	171	56.4	84.7
W12×26	**37.2**	**92.8**	140	**58.3**	87.7	**3.61**	5.46	**5.33**	**14.9**	**204**	**56.1**	84.2
W10×30	36.6	91.3	137	56.6	85.1	3.08	4.61	4.84	16.1	170	63.0	94.5
W8×35	34.7	86.6	130	54.5	81.9	1.62	2.43	7.17	27.0	127	50.3	75.5
W14×22	**33.2**	**82.8**	125	**50.6**	76.1	**4.78**	7.27	**3.67**	**10.4**	**199**	**63.0**	94.5
W10×26	31.3	78.1	117	48.7	73.2	2.91	4.34	4.80	14.9	144	53.6	80.3
W8×31[f]	30.4	75.8	114	48.0	72.2	1.58	2.37	7.18	24.8	110	45.6	68.4
W12×22	**29.3**	**73.1**	110	**44.4**	66.7	**4.68**	7.06	**3.00**	**9.13**	**156**	**64.0**	95.9
W8×28	27.2	67.9	102	42.4	63.8	1.67	2.50	5.72	21.0	98.0	45.9	68.9
W10×22	**26.0**	**64.9**	97.5	**40.5**	60.9	**2.68**	4.02	**4.70**	**13.8**	**118**	**49.0**	73.4
W12×19	**24.7**	**61.6**	92.6	**37.2**	55.9	**4.27**	6.43	**2.90**	**8.61**	**130**	**57.3**	86.0
W8×24	23.1	57.6	86.6	36.5	54.9	1.60	2.40	5.69	18.9	82.7	38.9	58.3
W10×19	**21.6**	**53.9**	81.0	**32.8**	49.4	**3.18**	4.76	**3.09**	**9.73**	**96.3**	**51.0**	76.5
W8×21	20.4	50.9	76.5	31.8	47.8	1.85	2.77	4.45	14.8	75.3	41.4	62.1
W12×16	**20.1**	**50.1**	75.4	**29.9**	44.9	**3.80**	5.73	**2.73**	**8.05**	**103**	**52.8**	79.2
W10×17	18.7	46.7	70.1	28.3	42.5	2.98	4.47	2.98	9.16	81.9	48.5	72.7
W12×14[v]	**17.4**	**43.4**	65.3	**26.0**	39.1	**3.43**	5.17	**2.66**	**7.73**	**88.6**	**42.8**	64.3
W8×18	17.0	42.4	63.8	26.5	39.9	1.74	2.61	4.34	13.5	61.9	37.4	56.2
W10×15	16.0	39.9	60.0	24.1	36.2	2.75	4.14	2.86	8.61	68.9	46.0	68.9
W8×15	13.6	33.9	51.0	20.6	31.0	1.90	2.85	3.09	10.1	48.0	39.7	59.6
W10×12[f]	**12.6**	**31.2**	46.9	**19.0**	28.6	**2.36**	3.53	**2.87**	**8.05**	**53.8**	**37.5**	56.3
W8×13	11.4	28.4	42.8	17.3	26.0	1.76	2.67	2.98	9.27	39.6	36.8	55.1
W8×10[f]	**8.87**	**21.9**	32.9	**13.6**	20.5	**1.54**	2.30	**3.14**	**8.52**	**30.8**	**26.8**	40.2

ASD	LRFD
$\Omega_b = 1.67$	$\phi_b = 0.90$
$\Omega_v = 1.50$	$\phi_v = 1.00$

[f] Shape exceeds compact limit for flexure with F_y = 50 ksi; tabulated values have been adjusted accordingly.

[v] Shape does not meet the h/t_w limit for shear in AISC *Specification* Section G2.1(a) with F_y = 50 ksi; therefore, ϕ_v = 0.90 and Ω_v = 1.67.

Table 3-6 (continued)
Maximum Total Uniform Load, kips
W-Shapes

F_y = 50 ksi

W16

Shape		W16×									
		45		40		36		31		26[v]	
Design		ASD	LRFD	ASD	LRFD	ASD	LRFD	ASD	LRFD	ASD	LRFD
Span, ft	6					188	281	175	262	141	212
	7	222	333	195	293	182	274	154	231	126	189
	8	205	309	182	274	160	240	135	203	110	166
	9	183	274	162	243	142	213	120	180	98.0	147
	10	164	247	146	219	128	192	108	162	88.2	133
	11	149	224	132	199	116	175	98.0	147	80.2	121
	12	137	206	121	183	106	160	89.8	135	73.5	111
	13	126	190	112	168	98.3	148	82.9	125	67.9	102
	14	117	176	104	156	91.2	137	77.0	116	63.0	94.7
	15	110	165	97.1	146	85.2	128	71.9	108	58.8	88.4
	16	103	154	91.1	137	79.8	120	67.4	101	55.1	82.9
	17	96.6	145	85.7	129	75.1	113	63.4	95.3	51.9	78.0
	18	91.3	137	80.9	122	71.0	107	59.9	90.0	49.0	73.7
	19	86.5	130	76.7	115	67.2	101	56.7	85.3	46.4	69.8
	20	82.1	123	72.9	110	63.9	96.0	53.9	81.0	44.1	66.3
	21	78.2	118	69.4	104	60.8	91.4	51.3	77.1	42.0	63.1
	22	74.7	112	66.2	99.5	58.1	87.3	49.0	73.6	40.1	60.3
	23	71.4	107	63.4	95.2	55.5	83.5	46.9	70.4	38.4	57.7
	24	68.4	103	60.7	91.3	53.2	80.0	44.9	67.5	36.8	55.3
	25	65.7	98.8	58.3	87.6	51.1	76.8	43.1	64.8	35.3	53.0
	26	63.2	95.0	56.0	84.2	49.1	73.8	41.5	62.3	33.9	51.0
	27	60.8	91.4	54.0	81.1	47.3	71.1	39.9	60.0	32.7	49.1
	28	58.7	88.2	52.0	78.2	45.6	68.6	38.5	57.9	31.5	47.4
	29	56.6	85.1	50.2	75.5	44.0	66.2	37.2	55.9	30.4	45.7
	30	54.8	82.3	48.6	73.0	42.6	64.0	35.9	54.0	29.4	44.2
	31	53.0	79.6	47.0	70.6	41.2	61.9	34.8	52.3	28.5	42.8
	32	51.3	77.2	45.5	68.4	39.9	60.0	33.7	50.6	27.6	41.4
	33	49.8	74.8	44.2	66.4	38.7	58.2	32.7	49.1	26.7	40.2
	34	48.3	72.6	42.9	64.4	37.6	56.5	31.7	47.6	25.9	39.0
	35	46.9	70.5	41.6	62.6	36.5	54.9	30.8	46.3	25.2	37.9
	36	45.6	68.6	40.5	60.8	35.5	53.3	29.9	45.0	24.5	36.8
	37	44.4	66.7	39.4	59.2	34.5	51.9	29.1	43.8	23.8	35.8
	38	43.2	65.0	38.3	57.6	33.6	50.5	28.4	42.6	23.2	34.9
	39	42.1	63.3	37.4	56.2	32.8	49.2	27.6	41.5	22.6	34.0
	40	41.1	61.7	36.4	54.8						
Beam Properties											
W_c/Ω_b	$\phi_b W_c$, kip-ft	1640	2470	1460	2190	1280	1920	1080	1620	882	1330
M_p/Ω_b	$\phi_b M_p$, kip-ft	205	309	182	274	160	240	135	203	110	166
M_r/Ω_b	$\phi_b M_r$, kip-ft	127	191	113	170	98.7	148	82.4	124	67.1	101
BF/Ω_b	$\phi_b BF$, kips	7.12	10.8	6.67	10.0	6.24	9.36	6.86	10.3	5.93	8.98
V_n/Ω_v	$\phi_v V_n$, kips	111	167	97.6	146	93.8	141	87.5	131	70.5	106
Z_x, in.^{3}		82.3		73.0		64.0		54.0		44.2	
L_p, ft		5.55		5.55		5.37		4.13		3.96	
L_r, ft		16.5		15.9		15.2		11.8		11.2	

ASD	LRFD
$\Omega_b = 1.67$	$\phi_b = 0.90$
$\Omega_v = 1.50$	$\phi_v = 1.00$

[v] Shape does not meet the h/t_w limit for shear in AISC *Specification* Section G2.1(a) with F_y = 50 ksi; therefore, $\phi_v = 0.90$ and $\Omega_v = 1.67$.

Notes: For beams laterally unsupported, see Table 3-10.
Available strength tabulated above heavy line is limited by available shear strength.

W8

Table 4-1a (continued)
Available Strength in Axial Compression, kips
W-Shapes

$F_y = 50$ ksi

Shape		W8×											
lb/ft		67		58		48		40		35		31	
Design		P_n/Ω_c	$\phi_c P_n$	P_n/Ω_c	$\phi_c P_n$	P_n/Ω_c	$\phi_c P_n$	P_n/Ω_c	$\phi_c P_n$	P_n/Ω_c	$\phi_c P_n$	P_n/Ω_c	$\phi_c P_n$
		ASD	LRFD	ASD	LRFD	ASD	LRFD	ASD	LRFD	ASD	LRFD	ASD	LRFD
Effective length, L_c (ft), with respect to least radius of gyration, r_y	0	590	886	512	769	422	634	350	526	308	463	273	411
	6	542	815	470	706	387	581	320	481	281	423	249	374
	7	526	790	455	685	375	563	309	465	272	409	241	362
	8	508	763	439	660	361	543	298	448	262	394	232	348
	9	488	733	422	634	347	521	285	429	251	377	222	333
	10	467	701	403	606	331	497	272	409	239	359	211	317
	11	444	668	384	576	314	473	258	388	226	340	200	301
	12	421	633	363	546	297	447	243	366	213	321	189	283
	13	397	597	342	514	280	421	228	343	200	301	177	266
	14	373	560	321	482	262	394	213	321	187	281	165	248
	15	348	523	299	450	244	367	198	298	174	261	153	230
	16	324	487	278	418	226	340	183	275	160	241	141	212
	17	300	450	257	386	209	314	169	253	147	221	130	195
	18	276	415	236	355	192	288	154	232	135	203	118	178
	19	253	381	216	325	175	264	141	211	123	184	108	162
	20	231	347	197	296	159	239	127	191	111	166	97.2	146
	22	191	287	163	244	132	198	105	158	91.5	138	80.3	121
	24	160	241	137	205	111	166	88.2	133	76.9	116	67.5	101
	26	137	205	116	175	94.2	142	75.2	113	65.5	98.5	57.5	86.5
	28	118	177	100	151	81.2	122	64.8	97.4	56.5	84.9	49.6	74.5
	30	103	154	87.5	131	70.7	106	56.5	84.9	49.2	74.0	43.2	64.9
	32	90.3	136	76.9	116	62.2	93.5	49.6	74.6	43.3	65.0	38.0	57.1
	34	79.9	120	68.1	102	55.1	82.8	44.0	66.1				
Properties													
P_{wo}, kips		126	190	102	153	72.0	108	57.2	85.9	45.9	68.9	39.4	59.1
P_{wi}, kip/in.		19.0	28.5	17.0	25.5	13.3	20.0	12.0	18.0	10.3	15.5	9.50	14.3
P_{wb}, kips		507	761	363	546	174	262	127	192	81.1	122	63.0	94.7
P_{fb}, kips		164	246	123	185	87.8	132	58.7	88.2	45.9	68.9	35.4	53.2
L_p, ft		7.49		7.42		7.35		7.21		7.17		7.18	
L_r, ft		47.6		41.6		35.2		29.9		27.0		24.8	
A_g, in.^{2}		19.7		17.1		14.1		11.7		10.3		9.13	
I_x, in.^{4}		272		228		184		146		127		110	
I_y, in.^{4}		88.6		75.1		60.9		49.1		42.0		37.1	
r_y, in.		2.12		2.10		2.08		2.04		2.03		2.02	
r_x/r_y		1.75		1.74		1.74		1.73		1.73		1.72	
$P_{ex}L_c^2/10^4$, k-in.^{2}		7790		6530		5270		4180		3630		3150	
$P_{ey}L_c^2/10^4$, k-in.^{2}		2540		2150		1740		1410		1220		1060	

ASD	LRFD	Note: Heavy line indicates L_c/r_y equal to or greater than 200.
$\Omega_c = 1.67$	$\phi_c = 0.90$	

Table 4-4 (continued)
Available Strength in Axial Compression, kips

Square HSS

HSS6

$F_y = 50$ ksi

Shape	HSS6×6×									
	3/8		5/16		1/4		3/16		1/8[c]	
t_{des}, in.	0.349		0.291		0.233		0.174		0.116	
lb/ft	27.48		23.34		19.02		14.53		9.86	
Design	P_n/Ω_c	$\phi_c P_n$	P_n/Ω_c	$\phi_c P_n$	P_n/Ω_c	$\phi_c P_n$	P_n/Ω_c	$\phi_c P_n$	P_n/Ω_c	$\phi_c P_n$
	ASD	LRFD	ASD	LRFD	ASD	LRFD	ASD	LRFD	ASD	LRFD
Effective length, L_c (ft), with respect to least radius of gyration, r_y: 0	227	341	193	289	157	236	119	179	63.1	94.8
6	211	317	179	270	146	220	111	167	60.5	90.9
7	206	309	175	263	143	215	109	163	59.6	89.5
8	199	300	170	255	139	208	106	159	58.5	87.9
9	193	289	164	247	134	202	102	154	57.3	86.2
10	185	279	158	238	129	195	98.8	148	56.1	84.3
11	178	267	152	228	124	187	95.0	143	54.7	82.2
12	170	255	145	218	119	179	91.0	137	53.2	80.0
13	161	242	138	207	113	170	86.8	130	51.6	77.6
14	153	229	131	197	108	162	82.5	124	50.0	75.1
15	144	216	123	186	102	153	78.2	117	48.2	72.5
16	135	203	116	175	95.9	144	73.7	111	46.4	69.8
17	126	190	109	164	90.0	135	69.3	104	44.5	67.0
18	118	177	102	153	84.1	126	64.9	97.6	42.6	64.1
19	109	164	94.4	142	78.4	118	60.6	91.0	40.7	61.1
20	101	152	87.4	131	72.7	109	56.3	84.6	38.7	58.1
21	92.9	140	80.6	121	67.2	101	52.1	78.4	35.9	53.9
22	85.0	128	74.0	111	61.9	93.0	48.1	72.3	33.1	49.8
23	77.8	117	67.7	102	56.6	85.1	44.1	66.3	30.4	45.7
24	71.4	107	62.2	93.5	52.0	78.1	40.5	60.9	27.9	42.0
25	65.8	98.9	57.3	86.1	47.9	72.0	37.3	56.1	25.8	38.7
26	60.8	91.4	53.0	79.6	44.3	66.6	34.5	51.9	23.8	35.8
27	56.4	84.8	49.1	73.8	41.1	61.7	32.0	48.1	22.1	33.2
28	52.5	78.8	45.7	68.7	38.2	57.4	29.8	44.7	20.5	30.9
29	48.9	73.5	42.6	64.0	35.6	53.5	27.7	41.7	19.1	28.8
30	45.7	68.7	39.8	59.8	33.3	50.0	25.9	39.0	17.9	26.9
32	40.2	60.4	35.0	52.6	29.2	44.0	22.8	34.2	15.7	23.6
34	35.6	53.5	31.0	46.6	25.9	38.9	20.2	30.3	13.9	20.9
36	31.7	47.7	27.6	41.5	23.1	34.7	18.0	27.1	12.4	18.7
38	28.5	42.8	24.8	37.3	20.7	31.2	16.2	24.3	11.1	16.8
Properties										
A_g, in.^{2}	7.58		6.43		5.24		3.98		2.70	
$I_x = I_y$, in.^{4}	39.5		34.3		28.6		22.3		15.5	
$r_x = r_y$, in.	2.28		2.31		2.34		2.37		2.39	

ASD	LRFD
$\Omega_c = 1.67$	$\phi_c = 0.90$

[c] Shape is slender for compression with $F_y = 50$ ksi; tabulated values have been adjusted accordingly.

F_y = 50 ksi

Table 4-4 (continued)
Available Strength in Axial Compression, kips
Square HSS

HSS5½–HSS5

Shape	HSS5½×5½×										HSS5×5×	
	3/8		5/16		1/4		3/16		1/8[c]		1/2	
t_{des}, in.	0.349		0.291		0.233		0.174		0.116		0.465	
lb/ft	24.93		21.21		17.32		13.25		9.01		28.43	
Design	P_n/Ω_c	$\phi_c P_n$	P_n/Ω_c	$\phi_c P_n$	P_n/Ω_c	$\phi_c P_n$	P_n/Ω_c	$\phi_c P_n$	P_n/Ω_c	$\phi_c P_n$	P_n/Ω_c	$\phi_c P_n$
	ASD	LRFD	ASD	LRFD	ASD	LRFD	ASD	LRFD	ASD	LRFD	ASD	LRFD
Effective length, L_c (ft), with respect to least radius of gyration, r_y												
0	206	310	175	263	143	215	109	163	61.4	92.3	236	355
6	189	284	161	242	131	197	100	151	58.4	87.7	210	316
7	183	275	156	234	127	192	97.3	146	57.3	86.1	202	303
8	176	265	151	226	123	185	94.1	141	56.1	84.3	192	289
9	169	254	145	217	118	178	90.5	136	54.7	82.2	182	274
10	161	243	138	208	113	170	86.7	130	53.2	80.0	172	258
11	153	231	132	198	108	162	82.7	124	51.6	77.6	161	241
12	145	218	125	187	102	154	78.5	118	49.9	75.0	149	224
13	137	205	117	177	96.5	145	74.2	112	48.1	72.2	138	207
14	128	192	110	166	90.6	136	69.8	105	46.2	69.4	127	190
15	119	179	103	155	84.7	127	65.4	98.3	44.2	66.4	115	173
16	110	166	95.6	144	78.8	118	61.0	91.7	42.0	63.1	105	157
17	102	153	88.4	133	73.0	110	56.6	85.1	39.1	58.7	94.2	142
18	93.6	141	81.4	122	67.3	101	52.3	78.6	36.2	54.4	84.1	126
19	85.6	129	74.6	112	61.8	92.9	48.1	72.3	33.3	50.1	75.5	113
20	77.7	117	68.0	102	56.4	84.8	44.1	66.2	30.6	46.0	68.1	102
21	70.5	106	61.6	92.7	51.2	77.0	40.1	60.2	27.9	42.0	61.8	92.9
22	64.2	96.5	56.2	84.4	46.7	70.1	36.5	54.9	25.4	38.2	56.3	84.6
23	58.7	88.3	51.4	77.2	42.7	64.2	33.4	50.2	23.3	35.0	51.5	77.4
24	53.9	81.1	47.2	70.9	39.2	58.9	30.7	46.1	21.4	32.1	47.3	71.1
25	49.7	74.7	43.5	65.4	36.1	54.3	28.3	42.5	19.7	29.6	43.6	65.5
26	46.0	69.1	40.2	60.4	33.4	50.2	26.2	39.3	18.2	27.4	40.3	60.6
27	42.6	64.1	37.3	56.0	31.0	46.6	24.2	36.4	16.9	25.4	37.4	56.2
28	39.6	59.6	34.7	52.1	28.8	43.3	22.5	33.9	15.7	23.6	34.8	52.2
29	36.9	55.5	32.3	48.6	26.9	40.4	21.0	31.6	14.6	22.0	32.4	48.7
30	34.5	51.9	30.2	45.4	25.1	37.7	19.6	29.5	13.7	20.6	30.3	45.5
Properties												
A_g, in.^{2}	6.88		5.85		4.77		3.63		2.46		7.88	
$I_x = I_y$, in.^{4}	29.7		25.9		21.7		17.0		11.8		26.0	
$r_x = r_y$, in.	2.08		2.11		2.13		2.16		2.19		1.82	

ASD	LRFD
$\Omega_c = 1.67$	$\phi_c = 0.90$

[c] Shape is slender for compression with F_y = 50 ksi; tabulated values have been adjusted accordingly.

Table 4-4 (continued)
Available Strength in Axial Compression, kips
Square HSS

HSS5–HSS4½

$F_y = 50$ ksi

Shape	HSS5×5×										HSS4½×4½×	
	3/8		5/16		1/4		3/16		1/8[c]		1/2	
t_{des}, in.	0.349		0.291		0.233		0.174		0.116		0.465	
lb/ft	22.37		19.08		15.62		11.97		8.16		25.03	
Design	P_n/Ω_c	$\phi_c P_n$	P_n/Ω_c	$\phi_c P_n$	P_n/Ω_c	$\phi_c P_n$	P_n/Ω_c	$\phi_c P_n$	P_n/Ω_c	$\phi_c P_n$	P_n/Ω_c	$\phi_c P_n$
	ASD	LRFD	ASD	LRFD	ASD	LRFD	ASD	LRFD	ASD	LRFD	ASD	LRFD
Effective length, L_c (ft), with respect to least radius of gyration, r_y												
0	185	278	157	237	129	193	98.2	148	59.8	89.9	208	313
1	184	277	157	236	128	193	97.9	147	59.7	89.7	207	311
2	183	275	156	234	127	191	97.1	146	59.4	89.2	205	308
3	180	271	153	231	126	189	95.8	144	58.8	88.4	201	302
4	176	265	150	226	123	185	94.0	141	58.1	87.3	195	293
5	172	258	146	220	120	180	91.7	138	57.2	86.0	188	283
6	166	250	142	213	116	175	89.0	134	56.1	84.3	180	270
7	160	240	137	205	112	168	85.9	129	54.8	82.3	171	256
8	153	229	131	196	107	161	82.4	124	53.3	80.1	160	241
9	145	218	124	187	102	154	78.7	118	51.7	77.7	150	225
10	137	206	118	177	97.0	146	74.7	112	49.9	75.1	139	208
11	129	193	111	166	91.5	137	70.5	106	48.0	72.2	127	191
12	120	180	103	156	85.7	129	66.2	99.5	45.5	68.4	116	174
13	111	167	96.2	145	79.8	120	61.8	92.9	42.6	64.0	105	157
14	103	154	88.9	134	74.0	111	57.4	86.3	39.6	59.6	93.9	141
15	94.0	141	81.7	123	68.2	102	53.0	79.7	36.7	55.2	83.4	125
16	85.6	129	74.6	112	62.4	93.8	48.7	73.2	33.8	50.8	73.5	110
17	75.5	116	67.8	102	56.9	85.5	44.5	66.8	31.0	46.5	65.1	97.8
18	69.6	105	61.2	91.9	51.5	77.4	40.4	60.7	28.2	42.4	58.0	87.2
19	62.5	93.9	54.9	82.5	46.3	69.6	36.4	54.8	25.5	38.4	52.1	78.3
20	56.4	84.8	49.6	74.5	41.8	62.8	32.9	49.4	23.0	34.6	47.0	70.7
21	51.2	76.9	44.9	67.6	37.9	57.0	29.8	44.8	20.9	31.4	42.6	64.1
22	46.6	70.0	41.0	61.5	34.5	51.9	27.2	40.8	19.0	28.6	38.9	58.4
23	42.6	64.1	37.5	56.3	31.6	47.5	24.9	37.4	17.4	26.2	35.5	53.4
24	39.2	58.9	34.4	51.7	29.0	43.6	22.8	34.3	16.0	24.1	32.6	49.1
25	36.1	54.2	31.7	47.7	26.7	40.2	21.0	31.6	14.7	22.2	30.1	45.2
26	33.4	50.2	29.3	44.1	24.7	37.2	19.5	29.2	13.6	20.5	27.8	41.8
27	30.9	46.5	27.2	40.9	22.9	34.5	18.0	27.1	12.6	19.0		
28	28.8	43.2	25.3	38.0	21.3	32.1	16.8	25.2	11.8	17.7		
29	26.8	40.3	23.6	35.4	19.9	29.9	15.6	23.5	11.0	16.5		
Properties												
A_g, in.^{2}	6.18		5.26		4.30		3.28		2.23		6.95	
$I_x = I_y$, in.^{4}	21.7		19.0		16.0		12.6		8.80		18.1	
$r_x = r_y$, in.	1.87		1.90		1.93		1.96		1.99		1.61	

ASD	LRFD
$\Omega_c = 1.67$	$\phi_c = 0.90$

[c] Shape is slender for compression with $F_y = 50$ ksi; tabulated values have been adjusted accordingly.

Note: Heavy line indicates L_c/r_y equal to or greater than 200.

Table 4-4 (continued)
Available Strength in Axial Compression, kips
Square HSS

F_y = 50 ksi

HSS4½–HSS4

Shape	HSS4½×4½×										HSS4×4×	
	3/8		5/16		1/4		3/16		1/8[c]		1/2	
t_{des}, in.	0.349		0.291		0.233		0.174		0.116		0.465	
lb/ft	19.82		16.96		13.91		10.70		7.31		21.63	
Design	P_n/Ω_c	$\phi_c P_n$	P_n/Ω_c	$\phi_c P_n$	P_n/Ω_c	$\phi_c P_n$	P_n/Ω_c	$\phi_c P_n$	P_n/Ω_c	$\phi_c P_n$	P_n/Ω_c	$\phi_c P_n$
	ASD	LRFD	ASD	LRFD	ASD	LRFD	ASD	LRFD	ASD	LRFD	ASD	LRFD
Effective length, L_c (ft), with respect to least radius of gyration, r_y: 0	164	247	140	211	115	173	87.7	132	57.7	86.8	180	271
1	163	246	140	210	115	172	87.4	131	57.6	86.6	179	269
2	162	243	138	208	113	170	86.5	130	57.2	86.0	176	265
3	159	238	136	204	111	167	85.1	128	56.6	85.0	172	258
4	154	232	132	199	109	163	83.0	125	55.6	83.6	166	249
5	149	224	128	192	105	158	80.5	121	54.5	81.9	158	237
6	143	215	123	185	101	152	77.5	117	53.1	79.8	149	224
7	136	205	117	176	96.8	145	74.1	111	50.9	76.5	139	209
8	129	194	111	167	91.8	138	70.4	106	48.4	72.8	128	193
9	121	182	104	157	86.5	130	66.4	99.8	45.7	68.8	117	176
10	112	169	97.3	146	80.9	122	62.2	93.5	43.0	64.6	106	160
11	104	156	90.2	136	75.1	113	57.9	87.0	40.1	60.2	95.0	143
12	95.3	143	82.9	125	69.3	104	53.5	80.4	37.1	55.8	84.1	126
13	86.7	130	75.7	114	63.4	95.4	49.1	73.7	34.1	51.3	73.6	111
14	78.3	118	68.6	103	57.7	86.7	44.7	67.2	31.2	46.9	63.7	95.8
15	70.2	105	61.7	92.8	52.1	78.3	40.5	60.8	28.4	42.6	55.5	83.5
16	62.3	93.7	55.1	82.9	46.7	70.2	36.4	54.7	25.6	38.4	48.8	73.3
17	55.2	83.0	48.8	73.4	41.5	62.4	32.4	48.7	22.9	34.4	43.2	65.0
18	49.2	74.0	43.6	65.5	37.0	55.6	28.9	43.4	20.4	30.7	38.6	58.0
19	44.2	66.4	39.1	58.8	33.2	49.9	25.9	39.0	18.3	27.5	34.6	52.0
20	39.9	59.9	35.3	53.0	30.0	45.1	23.4	35.2	16.5	24.9	31.2	46.9
21	36.2	54.4	32.0	48.1	27.2	40.9	21.2	31.9	15.0	22.5	28.3	42.6
22	33.0	49.5	29.2	43.8	24.8	37.3	19.4	29.1	13.7	20.5	25.8	38.8
23	30.2	45.3	26.7	40.1	22.7	34.1	17.7	26.6	12.5	18.8	23.6	35.5
24	27.7	41.6	24.5	36.8	20.8	31.3	16.3	24.4	11.5	17.3		
25	25.5	38.4	22.6	34.0	19.2	28.8	15.0	22.5	10.6	15.9		
26	23.6	35.5	20.9	31.4	17.7	26.7	13.9	20.8	9.78	14.7		
27	21.9	32.9	19.4	29.1	16.5	24.7	12.8	19.3	9.07	13.6		
28			18.0	27.1	15.3	23.0	11.9	18.0	8.44	12.7		
29							11.1	16.7	7.86	11.8		
Properties												
A_g, in.²	5.48		4.68		3.84		2.93		2.00		6.02	
$I_x = I_y$, in.⁴	15.3		13.5		11.4		9.02		6.35		11.9	
$r_x = r_y$, in.	1.67		1.70		1.73		1.75		1.78		1.41	

ASD	LRFD
$\Omega_c = 1.67$	$\phi_c = 0.90$

[c] Shape is slender for compression with F_y = 50 ksi; tabulated values have been adjusted accordingly.

Note: Heavy line indicates L_c/r_y equal to or greater than 200.

Table 4-14 (continued)
Available Critical Stress for Compression Members

	$F_y = 35$ ksi		$F_y = 36$ ksi		$F_y = 46$ ksi		$F_y = 50$ ksi		$F_y = 65$ ksi		$F_y = 70$ ksi	
$\frac{L_c}{r}$	F_{cr}/Ω_c	$\phi_c F_{cr}$	F_{cr}/Ω_c	$\phi_c F_{cr}$	F_{cr}/Ω_c	$\phi_c F_{cr}$	F_{cr}/Ω_c	$\phi_c F_{cr}$	F_{cr}/Ω_c	$\phi_c F_{cr}$	F_{cr}/Ω_c	$\phi_c F_{cr}$
	ksi	ksi	ksi	ksi	ksi	ksi	ksi	ksi	ksi	ksi	ksi	ksi
	ASD	LRFD	ASD	LRFD	ASD	LRFD	ASD	LRFD	ASD	LRFD	ASD	LRFD
41	19.2	28.9	19.7	29.7	24.6	37.0	26.5	39.8	33.2	49.9	35.3	53.0
42	19.2	28.8	19.6	29.5	24.5	36.8	26.3	39.5	32.9	49.5	35.0	52.6
43	19.1	28.7	19.6	29.4	24.3	36.6	26.2	39.3	32.6	49.1	34.7	52.1
44	19.0	28.5	19.5	29.3	24.2	36.3	26.0	39.1	32.4	48.7	34.4	51.7
45	18.9	28.4	19.4	29.1	24.0	36.1	25.8	38.8	32.1	48.3	34.1	51.2
46	18.8	28.3	19.3	29.0	23.9	35.9	25.6	38.5	31.8	47.8	33.8	50.7
47	18.7	28.1	19.2	28.9	23.8	35.7	25.5	38.3	31.6	47.4	33.4	50.3
48	18.6	28.0	19.1	28.7	23.6	35.4	25.3	38.0	31.3	47.0	33.1	49.8
49	18.5	27.9	19.0	28.5	23.4	35.2	25.1	37.7	31.0	46.6	32.8	49.3
50	18.4	27.7	18.9	28.4	23.3	35.0	24.9	37.5	30.7	46.1	32.5	48.8
51	18.3	27.6	18.8	28.3	23.1	34.8	24.8	37.2	30.4	45.7	32.1	48.3
52	18.3	27.4	18.7	28.1	23.0	34.5	24.6	36.9	30.1	45.2	31.8	47.8
53	18.2	27.3	18.6	28.0	22.8	34.3	24.4	36.7	29.8	44.8	31.4	47.3
54	18.1	27.1	18.5	27.8	22.6	34.0	24.2	36.4	29.5	44.3	31.1	46.7
55	18.0	27.0	18.4	27.6	22.5	33.8	24.0	36.1	29.2	43.9	30.8	46.2
56	17.9	26.8	18.3	27.5	22.3	33.5	23.8	35.8	28.9	43.4	30.4	45.7
57	17.7	26.7	18.2	27.3	22.1	33.3	23.6	35.5	28.6	43.0	30.1	45.2
58	17.6	26.5	18.1	27.1	22.0	33.0	23.4	35.2	28.3	42.5	29.7	44.6
59	17.5	26.4	17.9	27.0	21.8	32.8	23.2	34.9	28.0	42.0	29.4	44.1
60	17.4	26.2	17.8	26.8	21.6	32.5	23.0	34.6	27.6	41.5	29.0	43.6
61	17.3	26.0	17.7	26.6	21.4	32.2	22.8	34.3	27.3	41.1	28.6	43.0
62	17.2	25.9	17.6	26.5	21.3	32.0	22.6	34.0	27.0	40.6	28.3	42.5
63	17.1	25.7	17.5	26.3	21.1	31.7	22.4	33.7	26.7	40.1	27.9	42.0
64	17.0	25.5	17.4	26.1	20.9	31.4	22.2	33.4	26.4	39.6	27.6	41.4
65	16.9	25.4	17.3	25.9	20.7	31.2	22.0	33.0	26.0	39.2	27.2	40.9
66	16.8	25.2	17.1	25.8	20.5	30.9	21.8	32.7	25.7	38.7	26.8	40.3
67	16.7	25.0	17.0	25.6	20.4	30.6	21.6	32.4	25.4	38.2	26.5	39.8
68	16.5	24.9	16.9	25.4	20.2	30.3	21.4	32.1	25.1	37.7	26.1	39.2
69	16.4	24.7	16.8	25.2	20.0	30.1	21.1	31.8	24.8	37.2	25.7	38.7
70	16.3	24.5	16.7	25.0	19.8	29.8	20.9	31.4	24.4	36.7	25.4	38.2
71	16.2	24.3	16.5	24.8	19.6	29.5	20.7	31.1	24.1	36.2	25.0	37.6
72	16.1	24.2	16.4	24.7	19.4	29.2	20.5	30.8	23.8	35.7	24.7	37.1
73	16.0	24.0	16.3	24.5	19.2	28.9	20.3	30.5	23.5	35.3	24.3	36.5
74	15.8	23.8	16.2	24.3	19.1	28.6	20.1	30.2	23.1	34.8	23.9	36.0
75	15.7	23.6	16.0	24.1	18.9	28.4	19.8	29.8	22.8	34.3	23.6	35.4
76	15.6	23.4	15.9	23.9	18.7	28.1	19.6	29.5	22.5	33.8	23.2	34.9
77	15.5	23.3	15.8	23.7	18.5	27.8	19.4	29.2	22.2	33.3	22.8	34.3
78	15.4	23.1	15.6	23.5	18.3	27.5	19.2	28.8	21.8	32.8	22.5	33.8
79	15.2	22.9	15.5	23.3	18.1	27.2	19.0	28.5	21.5	32.3	22.1	33.3
80	15.1	22.7	15.4	23.1	17.9	26.9	18.8	28.2	21.2	31.8	21.8	32.7

ASD	LRFD
$\Omega_c = 1.67$	$\phi_c = 0.90$

Appendix 2

The National Design Specification (NDS) Package for Wood Construction

The 2012 Edition of the National Design Specification (NDS) Package for Wood Construction is referenced in this text and contains guidelines for the design and use of wood products. It consists of the following four documents:

1. ASD/LRFD NDS—National Design Specification for Wood Construction with Commentary

This document provides guidelines, formulae, and charts for guidance on the design of structural members under various load cases and combinations. It consists of the following parts:

1—General Requirements for Structural Design
2—Design Values for Structural Members
3—Design Provisions and Equations
4—Sawn Lumber
5—Structural Glued Laminated Timber
6—Round Timber Poles and Piles
7—Prefabricated Wood I-Joists
8—Structural Composite Lumber
9—Wood Structural Panels
10—Mechanical Connections
11—Dowel-Type Fasteners
12—Split Ring and Shear Plate Connectors
13—Timber Rivets
14—Shear Walls and Diaphragms
15—Special Loading Conditions
16—Fire Design of Wood Members

2. NDS Supplement—Design Values for Wood Construction

This document provides design properties such as area, moment of inertia, and section modulus for various structural sections commonly used in design. It consists of the following parts:

1—Sawn Lumber Grading Agencies

2—Species Combinations

3—Section Properties

4—Reference Design Values

3. ASD/LRFD Manual for Engineered Wood Construction

This document provides additional guidance for the design of engineered structural products. Similar to the ASD/LRFD NDS, it consists of following parts:

M 1—General Requirements for Structural Design

M 2—Design Values for Structural Members

M 3—Design Provisions and Equations

M 4—Sawn Lumber

M 5—Structural Glued Laminated Timber

M 6—Round Timber Poles and Piles

M 7—Prefabricated Wood I-Joists

M 8—Structural Composite Lumber

M 9—Wood Structural Panels

M 10—Mechanical Connections

M 11—Dowel-type Fasteners

M 12—Split Ring and Shear Plate Connectors

M 13—Timber Rivets

M 14—Shear Walls and Diaphragms

M 15—Special Loading Conditions

M 16—Fire Design

4. ASD/LRFD Wind and Seismic—Special Design Provisions for Wind and Seismic With Commentary

This document provides guidance in the design of members for resisting wind and seismic forces. It consists of the following parts:

1—Designer Flowchart

2—General Design Requirements

3—Members and Connections

4—Lateral Force Resisting Systems

The following tables and excerpts from the NDS Supplement are included here for use in the Chapter 8 Case Study for Design in Sawn Wood:

Table 1B Section Properties of Standard Dressed (S4S) Sawn Lumber

This table displays sectional properties for various nominal sizes of sawn lumber displaying dressed size, area, section modulus, moment of inertia, and weights per linear foot.

Table 4A Reference Design Values for Visually Graded Dimension Lumber (2″–4″ thick)

This table displays design properties for various grades of dimension lumber displaying information such as bending, tension, shear, and compression values (stresses), as well as modulus of elasticity.

Table 4D Reference Design Values for Visually Graded Timbers (5″ × 5″ and larger)

This table displays the same design properties as Table 4A for sections 5″ × 5″ or larger.

Table 1B Section Properties of Standard Dressed (S4S) Sawn Lumber

Nominal Size b x d	Standard Dressed Size (S4S) b x d in. x in.	Area of Section A in.^{2}	X-X AXIS Section Modulus S_{xx} in.^{3}	X-X AXIS Moment of Inertia I_{xx} in.^{4}	Y-Y AXIS Section Modulus S_{yy} in.^{3}	Y-Y AXIS Moment of Inertia I_{yy} in.^{4}	Approximate weight in pounds per linear foot (lbs/ft) of piece when density of wood equals: 25 lbs/ft^3	30 lbs/ft^3	35 lbs/ft^3	40 lbs/ft^3	45 lbs/ft^3	50 lbs/ft^3
Boards[1]												
1 x 3	3/4 x 2-1/2	1.875	0.781	0.977	0.234	0.088	0.326	0.391	0.456	0.521	0.586	0.651
1 x 4	3/4 x 3-1/2	2.625	1.531	2.680	0.328	0.123	0.456	0.547	0.638	0.729	0.820	0.911
1 x 6	3/4 x 5-1/2	4.125	3.781	10.40	0.516	0.193	0.716	0.859	1.003	1.146	1.289	1.432
1 x 8	3/4 x 7-1/4	5.438	6.570	23.82	0.680	0.255	0.944	1.133	1.322	1.510	1.699	1.888
1 x 10	3/4 x 9-1/4	6.938	10.70	49.47	0.867	0.325	1.204	1.445	1.686	1.927	2.168	2.409
1 x 12	3/4 x 11-1/4	8.438	15.82	88.99	1.055	0.396	1.465	1.758	2.051	2.344	2.637	2.930
Dimension Lumber (see NDS 4.1.3.2) and Decking (see NDS 4.1.3.5)												
2 x 3	1-1/2 x 2-1/2	3.750	1.56	1.953	0.938	0.703	0.651	0.781	0.911	1.042	1.172	1.302
2 x 4	1-1/2 x 3-1/2	5.250	3.06	5.359	1.313	0.984	0.911	1.094	1.276	1.458	1.641	1.823
2 x 5	1-1/2 x 4-1/2	6.750	5.06	11.39	1.688	1.266	1.172	1.406	1.641	1.875	2.109	2.344
2 x 6	1-1/2 x 5-1/2	8.250	7.56	20.80	2.063	1.547	1.432	1.719	2.005	2.292	2.578	2.865
2 x 8	1-1/2 x 7-1/4	10.88	13.14	47.63	2.719	2.039	1.888	2.266	2.643	3.021	3.398	3.776
2 x 10	1-1/2 x 9-1/4	13.88	21.39	98.93	3.469	2.602	2.409	2.891	3.372	3.854	4.336	4.818
2 x 12	1-1/2 x 11-1/4	16.88	31.64	178.0	4.219	3.164	2.930	3.516	4.102	4.688	5.273	5.859
2 x 14	1-1/2 x 13-1/4	19.88	43.89	290.8	4.969	3.727	3.451	4.141	4.831	5.521	6.211	6.901
3 x 4	2-1/2 x 3-1/2	8.75	5.10	8.932	3.646	4.557	1.519	1.823	2.127	2.431	2.734	3.038
3 x 5	2-1/2 x 4-1/2	11.25	8.44	18.98	4.688	5.859	1.953	2.344	2.734	3.125	3.516	3.906
3 x 6	2-1/2 x 5-1/2	13.75	12.60	34.66	5.729	7.161	2.387	2.865	3.342	3.819	4.297	4.774
3 x 8	2-1/2 x 7-1/4	18.13	21.90	79.39	7.552	9.440	3.147	3.776	4.405	5.035	5.664	6.293
3 x 10	2-1/2 x 9-1/4	23.13	35.65	164.9	9.635	12.04	4.015	4.818	5.621	6.424	7.227	8.030
3 x 12	2-1/2 x 11-1/4	28.13	52.73	296.6	11.72	14.65	4.883	5.859	6.836	7.813	8.789	9.766
3 x 14	2-1/2 x 13-1/4	33.13	73.15	484.6	13.80	17.25	5.751	6.901	8.051	9.201	10.35	11.50
3 x 16	2-1/2 x 15-1/4	38.13	96.90	738.9	15.89	19.86	6.619	7.943	9.266	10.59	11.91	13.24
4 x 4	3-1/2 x 3-1/2	12.25	7.15	12.51	7.146	12.51	2.127	2.552	2.977	3.403	3.828	4.253
4 x 5	3-1/2 x 4-1/2	15.75	11.81	26.58	9.188	16.08	2.734	3.281	3.828	4.375	4.922	5.469
4 x 6	3-1/2 x 5-1/2	19.25	17.65	48.53	11.23	19.65	3.342	4.010	4.679	5.347	6.016	6.684
4 x 8	3-1/2 x 7-1/4	25.38	30.66	111.1	14.80	25.90	4.405	5.286	6.168	7.049	7.930	8.811
4 x 10	3-1/2 x 9-1/4	32.38	49.91	230.8	18.89	33.05	5.621	6.745	7.869	8.993	10.12	11.24
4 x 12	3-1/2 x 11-1/4	39.38	73.83	415.3	22.97	40.20	6.836	8.203	9.570	10.94	12.30	13.67
4 x 14	3-1/2 x 13-1/4	46.38	102.41	678.5	27.05	47.34	8.051	9.661	11.27	12.88	14.49	16.10
4 x 16	3-1/2 x 15-1/4	53.38	135.66	1034	31.14	54.49	9.266	11.12	12.97	14.83	16.68	18.53
Timbers (5" x 5" and larger)[2]												
Post and Timber (see NDS 4.1.3.4 and 4.1.5.3)												
5 x 5	4-1/2 x 4-1/2	20.25	15.19	34.17	15.19	34.17	3.516	4.219	4.922	5.625	6.328	7.031
6 x 6	5-1/2 x 5-1/2	30.25	27.73	76.26	27.73	76.26	5.252	6.302	7.352	8.403	9.453	10.50
6 x 8	5-1/2 x 7-1/2	41.25	51.56	193.4	37.81	104.0	7.161	8.594	10.03	11.46	12.89	14.32
8 x 8	7-1/2 x 7-1/2	56.25	70.31	263.7	70.31	263.7	9.766	11.72	13.67	15.63	17.58	19.53
8 x 10	7-1/2 x 9-1/2	71.25	112.8	535.9	89.06	334.0	12.37	14.84	17.32	19.79	22.27	24.74
10 x 10	9-1/2 x 9-1/2	90.25	142.9	678.8	142.9	678.8	15.67	18.80	21.94	25.07	28.20	31.34
10 x 12	9-1/2 x 11-1/2	109.3	209.4	1204	173.0	821.7	18.97	22.76	26.55	30.35	34.14	37.93
12 x 12	11-1/2 x 11-1/2	132.3	253.5	1458	253.5	1458	22.96	27.55	32.14	36.74	41.33	45.92
12 x 14	11-1/2 x 13-1/2	155.3	349.3	2358	297.6	1711	26.95	32.34	37.73	43.13	48.52	53.91
14 x 14	13-1/2 x 13-1/2	182.3	410.1	2768	410.1	2768	31.64	37.97	44.30	50.63	56.95	63.28
14 x 16	13-1/2 x 15-1/2	209.3	540.6	4189	470.8	3178	36.33	43.59	50.86	58.13	65.39	72.66
16 x 16	15-1/2 x 15-1/2	240.3	620.6	4810	620.6	4810	41.71	50.05	58.39	66.74	75.08	83.42
16 x 18	15-1/2 x 17-1/2	271.3	791.1	6923	700.7	5431	47.09	56.51	65.93	75.35	84.77	94.18
18 x 18	17-1/2 x 17-1/2	306.3	893.2	7816	893.2	7816	53.17	63.80	74.44	85.07	95.70	106.3
18 x 20	17-1/2 x 19-1/2	341.3	1109	10813	995.3	8709	59.24	71.09	82.94	94.79	106.6	118.5
20 x 20	19-1/2 x 19-1/2	380.3	1236	12049	1236	12049	66.02	79.22	92.4	105.6	118.8	132.0
20 x 22	19-1/2 x 21-1/2	419.3	1502	16150	1363	13285	72.79	87.34	101.9	116.5	131.0	145.6
22 x 22	21-1/2 x 21-1/2	462.3	1656	17806	1656	17806	80.25	96.30	112.4	128.4	144.5	160.5
22 x 24	21-1/2 x 23-1/2	505.3	1979	23252	1810	19463	87.72	105.3	122.8	140.3	157.9	175.4
24 x 24	23-1/2 x 23-1/2	552.3	2163	25415	2163	25415	95.88	115.1	134.2	153.4	172.6	191.8

Table 1B Section Properties of Standard Dressed (S4S) Sawn Lumber (Cont.)

Nominal Size b x d	Standard Dressed Size (S4S) b x d in. x in.	Area of Section A in.^{2}	X-X AXIS Section Modulus S_{xx} in.^{3}	X-X AXIS Moment of Inertia I_{xx} in.^{4}	Y-Y AXIS Section Modulus S_{yy} in.^{3}	Y-Y AXIS Moment of Inertia I_{yy} in.^{4}	Approximate weight in pounds per linear foot (lbs/ft) of piece when density of wood equals: 25 lbs/ft^3	30 lbs/ft^3	35 lbs/ft^3	40 lbs/ft^3	45 lbs/ft^3	50 lbs/ft^3
Beams & Stringers (see NDS 4.1.3.3 and 4.1.5.3)												
6 x 10	5-1/2 x 9-1/2	52.25	82.73	393.0	47.90	131.7	9.071	10.89	12.70	14.51	16.33	18.14
6 x 12	5-1/2 x 11-1/2	63.25	121.2	697.1	57.98	159.4	10.98	13.18	15.37	17.57	19.77	21.96
6 x 14	5-1/2 x 13-1/2	74.25	167.1	1128	68.06	187.2	12.89	15.47	18.05	20.63	23.20	25.78
6 x 16	5-1/2 x 15-1/2	85.25	220.2	1707	78.15	214.9	14.80	17.76	20.72	23.68	26.64	29.60
6 x 18	5-1/2 x 17-1/2	96.25	280.7	2456	88.23	242.6	16.71	20.05	23.39	26.74	30.08	33.42
6 x 20	5-1/2 x 19-1/2	107.3	348.6	3398	98.31	270.4	18.62	22.34	26.07	29.79	33.52	37.24
6 x 22	5-1/2 x 21-1/2	118.3	423.7	4555	108.4	298.1	20.53	24.64	28.74	32.85	36.95	41.06
6 x 24	5-1/2 x 23-1/2	129.3	506.2	5948	118.5	325.8	22.44	26.93	31.41	35.90	40.39	44.88
8 x 12	7-1/2 x 11-1/2	86.3	165.3	950.5	107.8	404.3	14.97	17.97	20.96	23.96	26.95	29.95
8 x 14	7-1/2 x 13-1/2	101.3	227.8	1538	126.6	474.6	17.58	21.09	24.61	28.13	31.64	35.16
8 x 16	7-1/2 x 15-1/2	116.3	300.3	2327	145.3	544.9	20.18	24.22	28.26	32.29	36.33	40.36
8 x 18	7-1/2 x 17-1/2	131.3	382.8	3350	164.1	615.2	22.79	27.34	31.90	36.46	41.02	45.57
8 x 20	7-1/2 x 19-1/2	146.3	475.3	4634	182.8	685.5	25.39	30.47	35.55	40.63	45.70	50.78
8 x 22	7-1/2 x 21-1/2	161.3	577.8	6211	201.6	755.9	27.99	33.59	39.19	44.79	50.39	55.99
8 x 24	7-1/2 x 23-1/2	176.3	690.3	8111	220.3	826.2	30.60	36.72	42.84	48.96	55.08	61.20
10 x 14	9-1/2 x 13-1/2	128.3	288.6	1948	203.1	964.5	22.27	26.72	31.17	35.63	40.08	44.53
10 x 16	9-1/2 x 15-1/2	147.3	380.4	2948	233.1	1107	25.56	30.68	35.79	40.90	46.02	51.13
10 x 18	9-1/2 x 17-1/2	166.3	484.9	4243	263.2	1250	28.86	34.64	40.41	46.18	51.95	57.73
10 x 20	9-1/2 x 19-1/2	185.3	602.1	5870	293.3	1393	32.16	38.59	45.03	51.46	57.89	64.32
10 x 22	9-1/2 x 21-1/2	204.3	731.9	7868	323.4	1536	35.46	42.55	49.64	56.74	63.83	70.92
10 x 24	9-1/2 x 23-1/2	223.3	874.4	10274	353.5	1679	38.76	46.51	54.26	62.01	69.77	77.52
12 x 16	11-1/2 x 15-1/2	178.3	460.5	3569	341.6	1964	30.95	37.14	43.32	49.51	55.70	61.89
12 x 18	11-1/2 x 17-1/2	201.3	587.0	5136	385.7	2218	34.94	41.93	48.91	55.90	62.89	69.88
12 x 20	11-1/2 x 19-1/2	224.3	728.8	7106	429.8	2471	38.93	46.72	54.51	62.29	70.08	77.86
12 x 22	11-1/2 x 21-1/2	247.3	886.0	9524	473.9	2725	42.93	51.51	60.10	68.68	77.27	85.85
12 x 24	11-1/2 x 23-1/2	270.3	1058	12437	518.0	2978	46.92	56.30	65.69	75.07	84.45	93.84
14 x 18	13-1/2 x 17-1/2	236.3	689.1	6029	531.6	3588	41.02	49.22	57.42	65.63	73.83	82.03
14 x 20	13-1/2 x 19-1/2	263.3	855.6	8342	592.3	3998	45.70	54.84	63.98	73.13	82.27	91.41
14 x 22	13-1/2 x 21-1/2	290.3	1040	11181	653.1	4408	50.39	60.47	70.55	80.63	90.70	100.8
14 x 24	13-1/2 x 23-1/2	317.3	1243	14600	713.8	4818	55.08	66.09	77.11	88.13	99.14	110.2
16 x 20	15-1/2 x 19-1/2	302.3	982.3	9578	780.8	6051	52.47	62.97	73.46	83.96	94.45	104.9
16 x 22	15-1/2 x 21-1/2	333.3	1194	12837	860.9	6672	57.86	69.43	81.00	92.57	104.1	115.7
16 x 24	15-1/2 x 23-1/2	364.3	1427	16763	941.0	7293	63.24	75.89	88.53	101.2	113.8	126.5
18 x 22	17-1/2 x 21-1/2	376.3	1348	14493	1097	9602	65.32	78.39	91.45	104.5	117.6	130.6
18 x 24	17-1/2 x 23-1/2	411.3	1611	18926	1199	10495	71.40	85.68	99.96	114.2	128.5	142.8
20 x 24	19-1/2 x 23-1/2	458.3	1795	21089	1489	14521	79.56	95.47	111.4	127.3	143.2	159.1

1. According to the Southern Pine Inspection Bureau's (SPIB) Standard Grading Rules for Southern Pine Lumber: Section 265 stress rated boards:
   - Industrial 55 (IND 55) shall be graded as per No. 1 dimension
   - Industrial 45 (IND 45) shall be graded as per No. 2 dimension
   - Industrial 26 (IND 26) shall be graded as per No. 3 dimension

   See Table 4B for Southern Pine dimension lumber design values.

2. Neither Redwood nor Southern Pine are classified as Beams and Stringers or Posts and Timbers.

(Continued)

Table 4A (Cont.) Reference Design Values for Visually Graded Dimension Lumber (2" - 4" thick)[1,2,3]

(All species except Southern Pine—see Table 4B) (Tabulated design values are for normal load duration and dry service conditions. See NDS 4.3 for a comprehensive description of design value adjustment factors.)

USE WITH TABLE 4A ADJUSTMENT FACTORS

Species and commercial grade	Size classification	Design values in pounds per square inch (psi)							Specific Gravity[4]	Grading Rules Agency
		Bending	Tension parallel to grain	Shear parallel to grain	Compression perpendicular to grain	Compression parallel to grain	Modulus of Elasticity			
		F_b	F_t	F_v	$F_{c\perp}$	F_c	E	E_{min}	G	
BEECH-BIRCH-HICKORY										
Select Structural	2" & wider	1,450	850	195	715	1,200	1,700,000	620,000	0.71	NELMA
No. 1	2" & wider	1,050	600	195	715	950	1,600,000	580,000	0.71	NELMA
No. 2	2" & wider	1,000	600	195	715	750	1,500,000	550,000	0.71	NELMA
No. 3	2" & wider	575	350	195	715	425	1,300,000	470,000	0.71	NELMA
Stud	2" & wider	775	450	195	715	475	1,300,000	470,000	0.71	NELMA
Construction	2" - 4" wide	1,150	675	195	715	1,000	1,400,000	510,000	0.71	NELMA
Standard	2" - 4" wide	650	375	195	715	775	1,300,000	470,000	0.71	NELMA
Utility	2" - 4" wide	300	175	195	715	500	1,200,000	440,000	0.71	NELMA
COAST SITKA SPRUCE										
Select Structural	2" & wider	1300	950	125	455	1200	1,700,000	620,000	0.43	NLGA
No. 1/ No. 2	2" & wider	925	550	125	455	1100	1,500,000	550,000	0.43	NLGA
No. 3	2" & wider	525	325	125	455	625	1,400,000	510,000	0.43	NLGA
Stud	2" & wider	725	450	125	455	675	1,400,000	510,000	0.43	NLGA
Construction	2" - 4" wide	1050	650	125	455	1300	1,400,000	510,000	0.43	NLGA
Standard	2" - 4" wide	600	350	125	455	1100	1,300,000	470,000	0.43	NLGA
Utility	2" - 4" wide	275	175	125	455	725	1,200,000	440,000	0.43	NLGA
COTTONWOOD										
Select Structural	2" & wider	875	525	125	320	775	1,200,000	440,000	0.41	NSLB
No. 1	2" & wider	625	375	125	320	625	1,200,000	440,000	0.41	NSLB
No. 2	2" & wider	625	350	125	320	475	1,100,000	400,000	0.41	NSLB
No. 3	2" & wider	350	200	125	320	275	1,000,000	370,000	0.41	NSLB
Stud	2" & wider	475	275	125	320	300	1,000,000	370,000	0.41	NSLB
Construction	2" - 4" wide	700	400	125	320	650	1,000,000	370,000	0.41	NSLB
Standard	2" - 4" wide	400	225	125	320	500	900,000	330,000	0.41	NSLB
Utility	2" - 4" wide	175	100	125	320	325	900,000	330,000	0.41	NSLB
DOUGLAS FIR-LARCH										
Select Structural	2" & wider	1,500	1,000	180	625	1,700	1,900,000	690,000	0.50	WCLIB WWPA
No. 1 & Btr	2" & wider	1,200	800	180	625	1,550	1,800,000	660,000	0.50	WCLIB WWPA
No. 1	2" & wider	1,000	675	180	625	1,500	1,700,000	620,000	0.50	WCLIB WWPA
No. 2	2" & wider	900	575	180	625	1,350	1,600,000	580,000	0.50	WCLIB WWPA
No. 3	2" & wider	525	325	180	625	775	1,400,000	510,000	0.50	WCLIB WWPA
Stud	2" & wider	700	450	180	625	850	1,400,000	510,000	0.50	WCLIB WWPA
Construction	2" - 4" wide	1,000	650	180	625	1,650	1,500,000	550,000	0.50	WCLIB WWPA
Standard	2" - 4" wide	575	375	180	625	1,400	1,400,000	510,000	0.50	WCLIB WWPA
Utility	2" - 4" wide	275	175	180	625	900	1,300,000	470,000	0.50	WCLIB WWPA
DOUGLAS FIR-LARCH (NORTH)										
Select Structural	2" & wider	1,350	825	180	625	1,900	1,900,000	690,000	0.49	NLGA
No. 1 & Btr	2" & wider	1,150	750	180	625	1,800	1,800,000	660,000	0.49	NLGA
No. 1/ No. 2	2" & wider	850	500	180	625	1,400	1,600,000	580,000	0.49	NLGA
No. 3	2" & wider	475	300	180	625	825	1,400,000	510,000	0.49	NLGA
Stud	2" & wider	650	400	180	625	900	1,400,000	510,000	0.49	NLGA
Construction	2" - 4" wide	950	575	180	625	1,800	1,500,000	550,000	0.49	NLGA
Standard	2" - 4" wide	525	325	180	625	1,450	1,400,000	510,000	0.49	NLGA
Utility	2" - 4" wide	250	150	180	625	950	1,300,000	470,000	0.49	NLGA
DOUGLAS FIR-SOUTH										
Select Structural	2" & wider	1,350	900	180	520	1,600	1,400,000	510,000	0.46	WWPA
No. 1	2" & wider	925	600	180	520	1,450	1,300,000	470,000	0.46	WWPA
No. 2	2" & wider	850	525	180	520	1,350	1,200,000	440,000	0.46	WWPA
No. 3	2" & wider	500	300	180	520	775	1,100,000	400,000	0.46	WWPA
Stud	2" & wider	675	425	180	520	850	1,100,000	400,000	0.46	WWPA
Construction	2" - 4" wide	975	600	180	520	1,650	1,200,000	440,000	0.46	WWPA
Standard	2" - 4" wide	550	350	180	520	1,400	1,100,000	400,000	0.46	WWPA
Utility	2" - 4" wide	250	150	180	520	900	1,000,000	370,000	0.46	WWPA

Table 4D Reference Design Values for Visually Graded Timbers (5" x 5" and larger)[1,3]

(Tabulated design values are for normal load duration and dry service conditions, unless specified otherwise. See NDS 4.3 for a comprehensive description of design value adjustment factors.)

USE WITH TABLE 4D ADJUSTMENT FACTORS

Species and commercial Grade	Size classification	Design values in pounds per square inch (psi): Bending F_b	Tension parallel to grain F_t	Shear parallel to grain F_v	Compression perpendicular to grain $F_{c\perp}$	Compression parallel to grain F_c	Modulus of Elasticity E	Modulus of Elasticity E_{min}	Specific Gravity[4] G	Grading Rules Agency
ALASKA CEDAR										
Select Structural	Beams and Stringers	1,400	675	155	525	925	1,200,000	440,000	0.47	WCLIB
No.1		1,150	475	155	525	775	1,200,000	440,000		
No.2		750	300	155	525	500	1,000,000	370,000		
Select Structural	Posts and Timbers	1,300	700	155	525	975	1,200,000	440,000		
No.1		1,050	575	155	525	850	1,200,000	440,000		
No.2		625	350	155	525	600	1,000,000	370,000		
BALDCYPRESS										
Select Structural	5"x5" and Larger	1,150	750	200	615	1,050	1,300,000	470,000	0.43	SPIB
No.1		1,000	675	200	615	925	1,300,000	470,000		
No.2		625	425	175	615	600	1,000,000	370,000		
BALSAM FIR										
Select Structural	Beams and Stringers	1,350	900	125	305	950	1,400,000	510,000	0.36	NELMA NSLB
No.1		1,100	750	125	305	800	1,400,000	510,000		
No.2		725	350	125	305	500	1,100,000	400,000		
Select Structural	Posts and Timbers	1,250	825	125	305	1,000	1,400,000	510,000		
No.1		1,000	675	125	305	875	1,400,000	510,000		
No.2		575	375	125	305	400	1,100,000	400,000		
BEECH-BIRCH-HICKORY										
Select Structural	Beams and Stringers	1,650	975	180	715	975	1,500,000	550,000	0.71	NELMA NSLB
No.1		1,400	700	180	715	825	1,500,000	550,000		
No.2		900	450	180	715	525	1,200,000	440,000		
Select Structural	Posts and Timbers	1,550	1,050	180	715	1,050	1,500,000	550,000		
No.1		1,250	850	180	715	900	1,500,000	550,000		
No.2		725	475	180	715	425	1,200,000	440,000		
COAST SITKA SPRUCE										
Select Structural	Beams and Stringers	1,150	675	115	455	775	1,500,000	550,000	0.43	NLGA
No.1		950	475	115	455	650	1,500,000	550,000		
No.2		625	325	115	455	425	1,200,000	440,000		
Select Structural	Posts and Timbers	1,100	725	115	455	825	1,500,000	550,000		
No.1		875	575	115	455	725	1,500,000	550,000		
No.2		525	350	115	455	500	1,200,000	440,000		
DOUGLAS FIR-LARCH										
Dense Select Structural	Beams and Stringers	1,900	1,100	170	730	1,300	1,700,000	620,000	0.50	WCLIB
Select Structural		1,600	950	170	625	1,100	1,600,000	580,000		
Dense No. 1		1,550	775	170	730	1,100	1,700,000	620,000		
No. 1		1,350	675	170	625	925	1,600,000	580,000		
No. 2		875	425	170	625	600	1,300,000	470,000		
Dense Select Structural	Posts and Timbers	1,750	1,150	170	730	1,350	1,700,000	620,000		
Select Structural		1,500	1,000	170	625	1,150	1,600,000	580,000		
Dense No. 1		1,400	950	170	730	1,200	1,700,000	620,000		
No. 1		1,200	825	170	625	1,000	1,600,000	580,000		
No. 2		750	475	170	625	700	1,300,000	470,000		
Dense Select Structural	Beams and Stringers	1,900	1,100	170	730	1,300	1,700,000	620,000	0.50	WWPA
Select Structural		1,600	950	170	625	1,100	1,600,000	580,000		
Dense No. 1		1,550	775	170	730	1,100	1,700,000	620,000		
No. 1		1,350	675	170	625	925	1,600,000	580,000		
No. 2 Dense		1,000	500	170	730	700	1,400,000	510,000		
No. 2		875	425	170	625	600	1,300,000	470,000		
Dense Select Structural	Posts and Timbers	1,750	1,150	170	730	1,350	1,700,000	620,000		
Select Structural		1,500	1,000	170	625	1,150	1,600,000	580,000		
Dense No. 1		1,400	950	170	730	1,200	1,700,000	620,000		
No. 1		1,200	825	170	625	1,000	1,600,000	580,000		
No. 2 Dense		850	550	170	730	825	1,400,000	510,000		
No. 2		750	475	170	625	700	1,300,000	470,000		

Appendix 3

ACI 318—Building Code Requirements for Structural Concrete

ACI Building Code Requirements for Structural Concrete (ACI-318–11) is referenced in this text and contains guidelines for the design and use of structural concrete. It consists of the following chapters and appendices:

Chapter 1—General Requirements
Chapter 2—Notation and Definitions
Chapter 3—Materials
Chapter 4—Durability Requirements
Chapter 5—Concrete Quality, Mixing, and Placing
Chapter 6—Formwork, Embedments, and Construction Joints
Chapter 7—Details of Reinforcement
Chapter 8—Analysis and Design—General Considerations
Chapter 9—Strength and Serviceability Requirements
Chapter 10—Flexure and Axial Loads
Chapter 11—Shear and Torsion
Chapter 12—Development and Splices of Reinforcement
Chapter 13—Two-Way Slab Systems
Chapter 14—Walls
Chapter 15—Footings
Chapter 16—Precast Concrete
Chapter 17—Composite Concrete Flexural Members
Chapter 18—Prestressed Concrete
Chapter 19—Shells and Folded Plate Members
Chapter 20—Strength Evaluation of Existing Structures
Chapter 21—Earthquake-Resistant Structures
Chapter 22—Structural Plain Concrete

Appendix A—Strut-and-Tie Models

Appendix B—Alternative Provisions for Reinforced and Prestressed Concrete Flexural and Compression Members

Appendix C—Alternative Load and Strength Reduction Factors

Appendix D—Anchoring to Concrete

Appendix E—Steel Reinforcement Information

Information in the text has been excerpted from ACI 318. The Reader is referred to the following sections for additional information on text references.

7.7—Concrete Protection for Reinforcement

This section provides minimum concrete clear cover for reinforcement for various exposures, members, and conditions.

9.5—Control of Deflections

This section provides minimum depths and thicknesses (in terms of span-to-depth ratios) to limit deflections for beams and one-way slabs.

10.5—Limits for Reinforcement of Flexural Members

This section provides guidelines for limits of reinforcement in flexural members.

10.9—Limits for Reinforcement of Compression Members

This section provides guidelines for limits of reinforcement in compression members.

10.10—Slenderness Effects in Compression Members

This section provides information on column classifications and slenderness ratios in concrete columns.

11.5.4—Spacing Limits for Shear Reinforcement

This section provides guidelines for the maximum spacing of shear reinforcement (i.e., stirrups) in members.

In addition to ACI 318–11, the *Reinforced Concrete Design Manual* provides supplemental information with tables and charts to provide assistance to professionals engaged in the design of reinforced concrete buildings and related structures. Table 1.6 (Flexure Design Aids) used in the Case Study is reproduced below.

1.6—Flexure design aids

Flexure 1: Flexural coefficients for rectangular beams with tension reinforcement; f_y = 60,000 psi

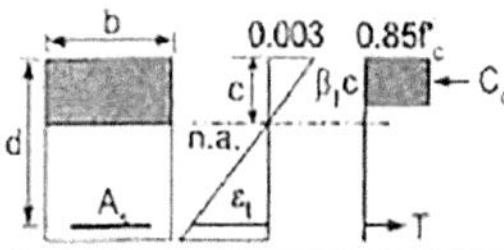

$\phi M_n \geq M_u$ $\qquad \phi M_n = \phi K_n bd^2/12{,}000$ $\qquad \rho = A_s/bd$

where M_n is in ft-kip; K_n is in psi; b and d are in inches.

f_y = 60,000 psi

f_c', psi:			3000		4000		5000		6000	
β_1:			0.85		0.85		0.80		0.75	
ρ_{min}:			0.0033		0.0033		0.0035		0.0039	
ε_t	ϕ	$\phi_{App\,C}$	ρ, %	ϕK_n, psi	ρ, %	ϕK_n, psi	ρ, %	ϕK_n, psi	ρ, %	ϕK_n, psi
0.20000	0.90	0.90	0.05	29	0.07	38	0.08	45	0.09	51
0.15000	0.90	0.90	0.07	38	0.09	51	0.11	60	0.13	67
0.10000	0.90	0.90	0.11	56	0.14	75	0.17	88	0.19	99
0.07500	0.90	0.90	0.14	74	0.19	98	0.22	116	0.25	130
0.05000	0.90	0.90	0.20	108	0.27	144	0.32	169	0.36	191
0.04000	0.90	0.90	0.25	132	0.34	176	0.40	208	0.44	234
0.03500	0.90	0.90	0.29	149	0.38	198	0.45	234	0.50	264
0.03000	0.90	0.90	0.33	170	0.44	227	0.52	268	0.58	302
0.02500	0.90	0.90	0.39	199	0.52	266	0.61	314	0.68	354
0.02000	0.90	0.90	0.47	240	0.63	320	0.74	378	0.83	427
0.01900	0.90	0.90	0.49	251	0.66	334	0.77	395	0.87	445
0.01800	0.90	0.90	0.52	262	0.69	349	0.81	412	0.91	465
0.01700	0.90	0.90	0.54	274	0.72	365	0.85	431	0.96	487
0.01600	0.90	0.90	0.57	287	0.76	383	0.89	453	1.01	511
0.01500	0.90	0.90	0.60	302	0.80	403	0.94	476	1.06	538
0.01400	0.90	0.90	0.64	318	0.85	425	1.00	502	1.13	567
0.01300	0.90	0.90	0.68	337	0.90	449	1.06	531	1.20	600
0.01250	0.90	0.90	0.70	347	0.93	462	1.10	546	1.23	618
0.01200	0.90	0.90	0.72	357	0.96	476	1.13	563	1.28	637
0.01150	0.90	0.90	0.75	368	1.00	491	1.17	581	1.32	657
0.01100	0.90	0.90	0.77	380	1.03	507	1.21	600	1.37	678
0.01050	0.90	0.90	0.80	393	1.07	523	1.26	620	1.42	701
0.01000	0.90	0.90	0.83	406	1.11	541	1.31	641	1.47	726
0.00950	0.90	0.90	0.87	420	1.16	561	1.36	664	1.53	752
0.00900	0.90	0.90	0.90	436	1.20	581	1.42	689	1.59	780
0.00870	0.90	0.90	0.93	446	1.24	594	1.45	704	1.63	798
0.00840	0.90	0.90	0.95	456	1.27	608	1.49	720	1.68	817
0.00810	0.90	0.90	0.98	467	1.30	622	1.53	738	1.72	836
0.00770	0.90	0.90	1.01	482	1.35	642	1.59	762	1.79	864
0.00740	0.90	0.90	1.04	494	1.39	658	1.63	781	1.84	886
0.00710	0.90	0.90	1.07	506	1.43	675	1.68	801	1.89	909
0.00680	0.90	0.90	1.11	519	1.47	693	1.73	822	1.95	933
0.00650	0.90	0.90	1.14	533	1.52	711	1.79	844	2.01	958
0.00620	0.90	0.90	1.18	548	1.57	731	1.85	868	2.08	985
0.00590	0.90	0.90	1.22	563	1.62	751	1.91	892	2.15	1014
0.00560	0.90	0.90	1.26	580	1.68	773	1.98	918	2.22	1044
0.00530	0.90	0.90	1.31	597	1.74	796	2.05	946	2.30	1076
0.00500	0.90	0.90	1.35	615	1.81	820	2.13	975	2.39	1109
0.00480	0.88	0.89	1.39	616	1.85	821	2.18	977	2.45	1112
0.00460	0.87	0.87	1.43	617	1.90	823	2.24	979	2.52	1115
0.00440	0.85	0.86	1.46	618	1.95	824	2.30	982	2.58	1118
0.00430	0.84	0.85	1.48	619	1.98	825	2.33	983	2.62	1119
0.00420	0.83	0.85	1.51	619	2.01	826	2.36	984	2.66	1121
0.00410	0.82	0.84	1.53	620	2.04	827	2.39	985	2.69	1122
0.00400	0.82	0.83	1.55	620	2.06	827	2.43	986	2.73	1124

Notes: The values of ρ above the rule are less than ρ_{min}. ϕK_n values are based on ϕ-factors provided in Chapter 9. When Appendix C values of ϕ are used, ϕK_n values in the transition zone may be up to 2.4% higher (more conservative).

DESIGN AID 1-6[1]

Maximum Number of Reinforcing Bars Permitted in a Single Layer

7.6

Assumptions:

- Grade 60 reinforcement (f_y = 60 ksi)
- Clear cover to the stirrups c_s = 1.5 in.
- ¾-in. aggregate
- No. 3 stirrups are used for No. 5 and No. 6 longitudinal bars, and No. 4 stirrups are used for No. 7 and larger bars

Bar Size	Beam Width (in.)												
	12	14	16	18	20	22	24	26	28	30	36	42	48
No. 4	5	6	8	9	10	12	13	14	16	17	21	25	29
No. 5	5	6	7	8	10	11	12	13	15	16	19	23	27
No. 6	4	6	7	8	9	10	11	12	14	15	18	22	25
No. 7	4	5	6	7	8	9	10	11	12	13	17	20	23
No. 8	4	5	6	7	8	9	10	11	12	13	16	19	22
No. 9	3	4	5	6	7	8	8	9	10	11	14	17	19
No. 10	3	4	4	5	6	7	8	8	9	10	12	15	17
No. 11	3	3	4	5	5	6	7	8	8	9	11	13	15

Maximum number of bars, n_{max}:

$$n_{max} = \frac{b_w - 2(c_s + d_s + r)}{(\text{Clear space}) + d_b} + 1$$

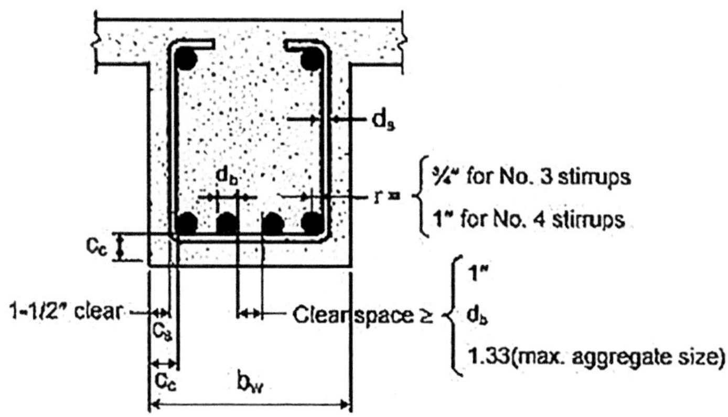

Appendix 4

Beam Diagrams and Formulae

The following are shear and moment diagrams, along with related formulae, for simple beams under various common loading conditions. A simple beam is a beam with simple supports (i.e., roller, friction-less surface, or pinned).

Simple Beam **Uniformly Distributed Load**	**Simple Beam** **Concentrated Load at Midspan**
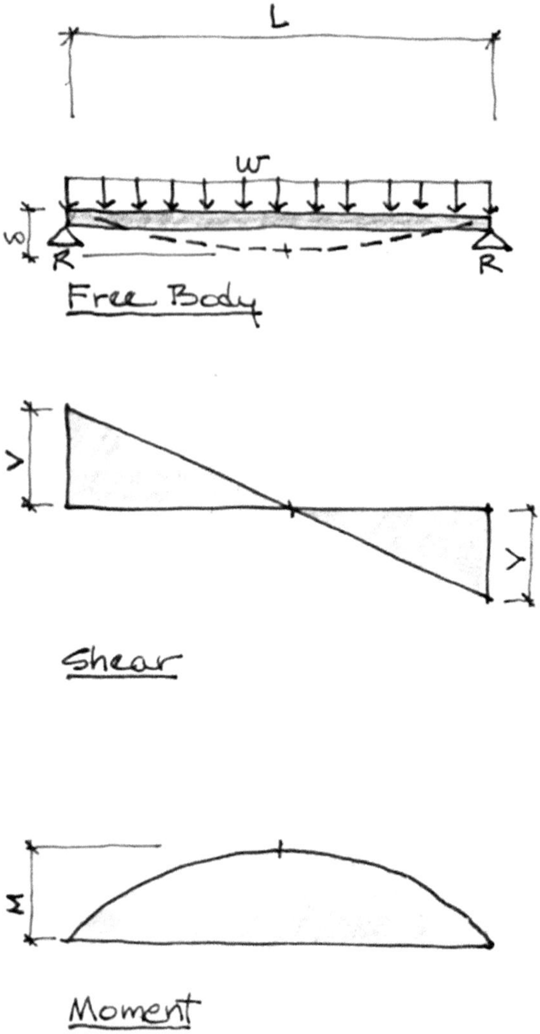	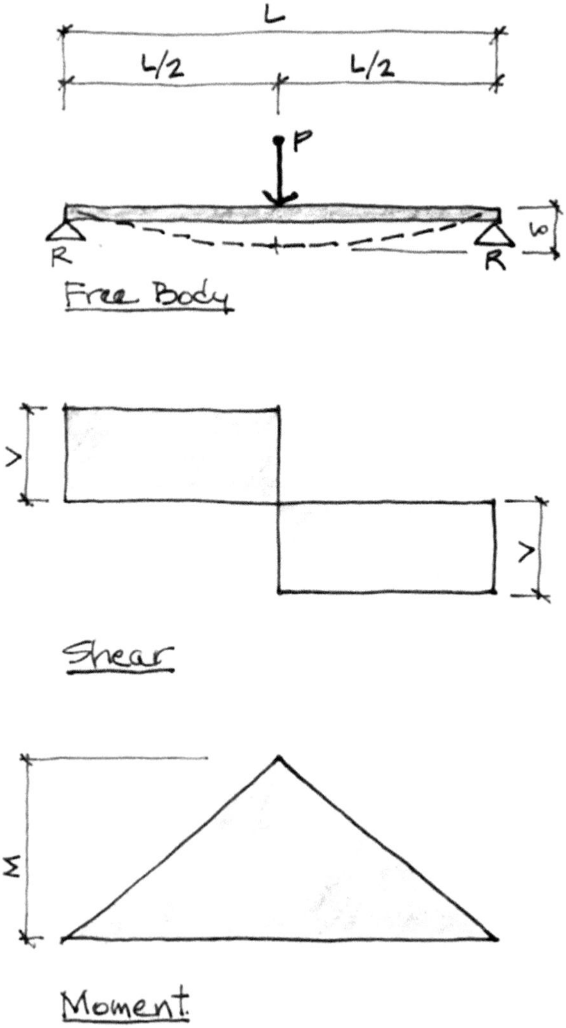
$R = wL/2$ $V = R$ $M = wL^2/8$ $\delta = 5wL^4/384EI$ or $\delta = 5(wL \times L^3)/384EI$	$R = P/2$ $V = R$ $M = PL/4$ $\delta = PL^3/48EI$

Figure A4.1

Concentrated Load at Any Point

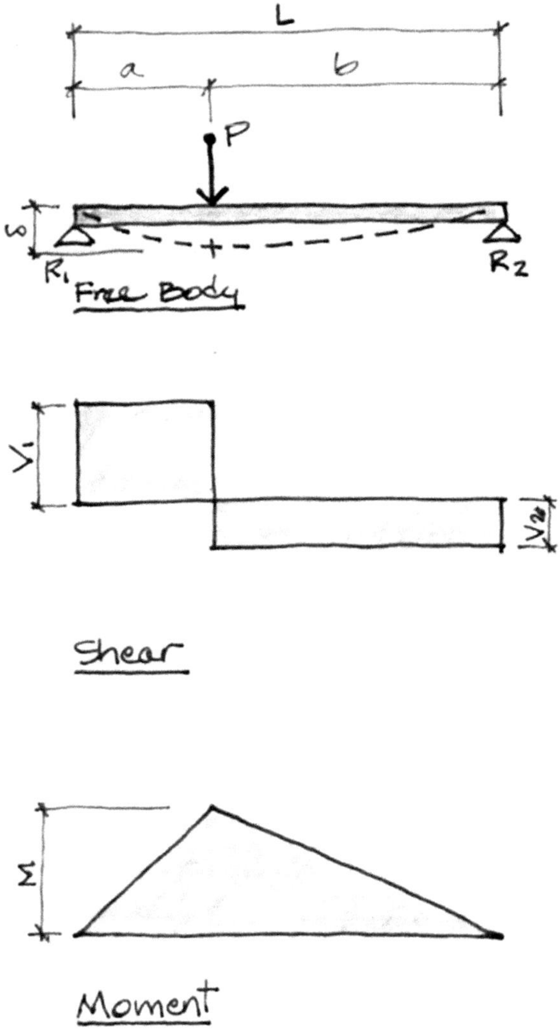

$R_1 = Pb/L$

$R_2 = Pa/L$

$V_1 = R_1$

$V_2 = R_2$

$M = Pab/L$

$\delta = Pa^2b^2/3EIL$

Two Equal Loads at Third Points

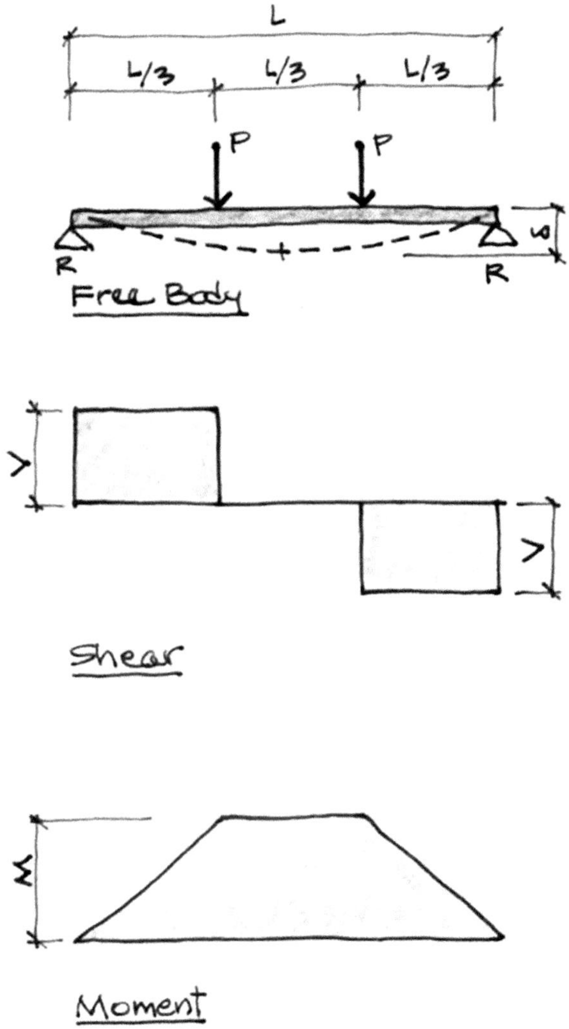

$R = P$

$V = R$

$M = PL/3$

$\delta = 0.036\ PL^3/EI$

Figure A4.2

Simple Beam
Three Equal Loads at Quarter Points

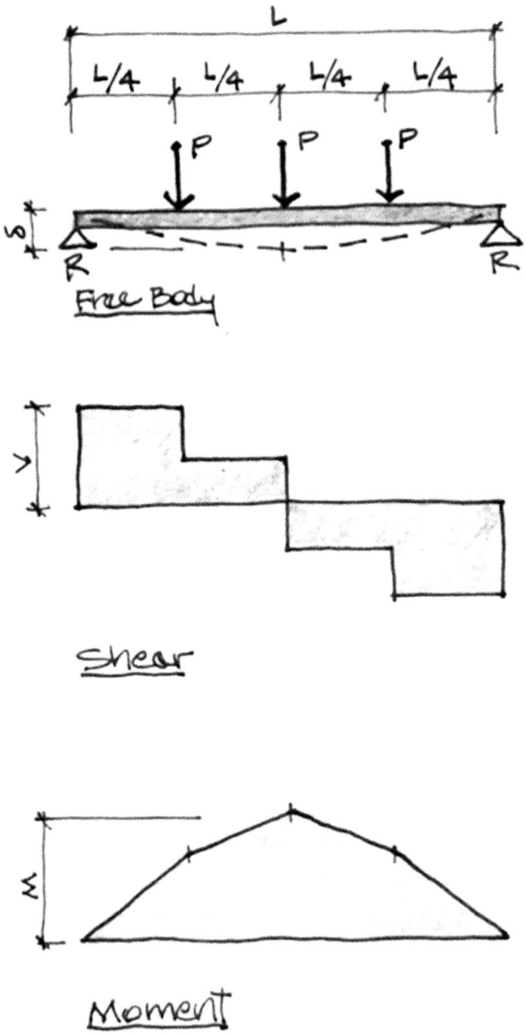

R = 3P/2
V = R
M = 0.5 PL
δ = 0.05 PL^3/EI

Simple Beam
Four Equal Loads at Fifth Points

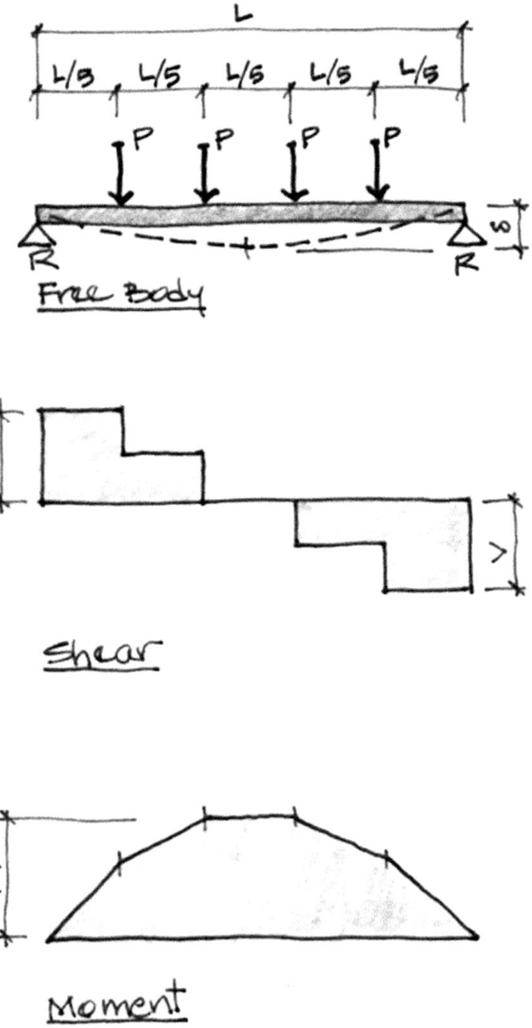

R = 4P/2
V = R
M = 0.6 PL
δ = 0.063 PL^3/EI

Figure A4.3

Table A4.1 Formula Coefficients for Symmetrically Loaded Beams

LOADING CASE	Coeff.	Support Type		
		Simple Both Ends	Fixed One End, Simple at Other	Fixed Both Ends

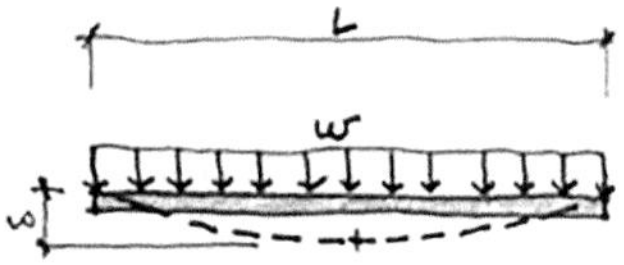	P = wL			
	a	0.125	0.070	0.042
	b	---	0.125	0.083
	c	0.500	0.375	---
	d	---	0.625	0.500
	e	0.013	0.005	0.003
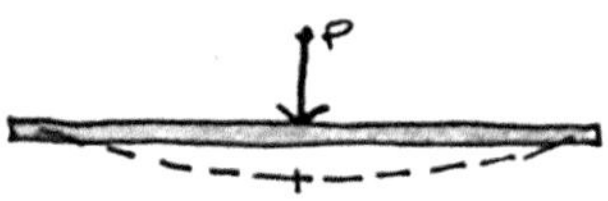	a	0.250	0.156	0.125
	b	---	0.188	0.125
	c	0.500	0.313	---
	d	---	0.688	0.500
	e	0.021	0.009	0.005
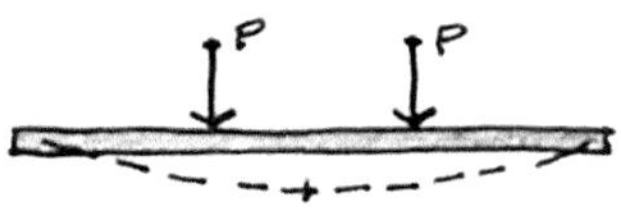	a	0.333	0.222	0.111
	b	---	0.333	0.222
	c	1.000	0.667	---
	d	---	1.333	1.000
	e	0.036	0.015	0.008
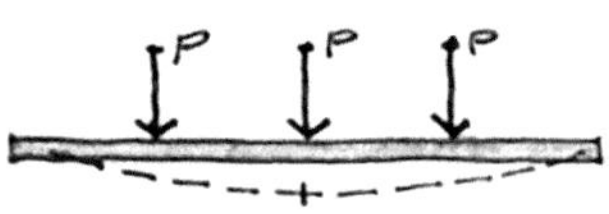	a	0.500	0.266	0.188
	b	---	0.469	0.313
	c	1.500	1.031	---
	d	---	1.969	1.500
	e	0.050	0.021	0.010
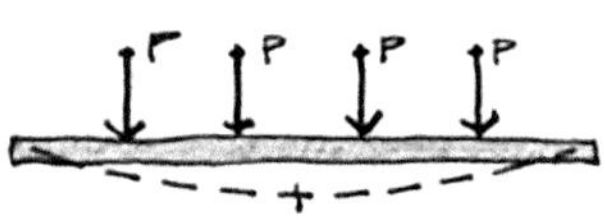	a	0.600	0.360	0.200
	b	---	0.600	0.400
	c	2.000	1.400	---
	d	---	2.600	2.000
	e	0.063	0.027	0.013

Maximum positive moment	= a PL
Maximum negative moment	= b PL
Simple end reaction	= c P
Fixed end reaction	= d P
Maximum deflection	= e PL^3/EI

Index

ACI 318–Building Code Requirements for Structural Concrete 5
adjusted design value *see* wood
adjustment factors *see* wood
Allowable Strength Design (ASD) 11, 13, 177
Allowable Stress Design 9–10, 13, 177
American Concrete Institute (ACI) 5, 213–216
American Institute of Steel Construction (AISC) 5, 189–204
American National Standards Institute (ANSI) 5
American Society for Testing and Materials (ASTM) 5
American Society of Civil Engineers (ASCE) 5
American Wood Council (AWC) 5
anisotropic 85, 177
ASCE 7, 7
ASD/LRFD comparison 82

balanced design *see* concrete
Beam Diagrams and Formulae 217–221
beam penetrations 40
beams/girders: buckling/bracing 39; design considerations (concrete 133–147; steel 48–51; wood 81–91); designing for (general) 35–40; *see also* concrete; steel; wood
beam theory (internal moment) 136–137
blast furnace 45
brittle 20, 178

capacity 9
codes 4–5
columns: buckling/crushing 42; design considerations (concrete 148–151; steel 51–53; wood 91–92); designing for (general) 41–43
compact steel sections 51, 178
concrete: admixtures 129, 178; balanced design 140, 178; control/expansion joints 131–132, 179, 180; cover 135, 179; cracks 143–144, 147; creep 179; curing 130, 179; deflection 147; design considerations (beams 133–147; columns 148–151); design example 152–174; fixity of connections 153, 180; hydration/curing 130, 179, 181; ingredients 129; materials and manufacture 128–129; singly and doubly reinforced beams 135; steel reinforcing 133; strength 131, 132; tension controlled sections 140, 187
critical load/stress (of a column) 41–42, 179
critical stress (of a steel column) 52
cross laminated timber (CLT) 95

dead load *see* loads
deflection 37–39, 179; in a concrete beam 147
deflection formulae 37–39, 218–221
deformation 179
design loads *see* loads
design methodologies 8–13, 179
ductility 16, 18, 179

effective depth (of a concrete beam) 135
effective length (of a column) 41, 180
elastic behavior 22–24, 25
elastic design 27, 180
elastic limit (for steel) 17–18, 180
elastic range 180; for concrete 21; for steel 17–18; for wood 20
elastic section modulus *see* section modulus
environmental load *see* loads
Euler's formula 42, 52, 180

factored load 8, 180
flexure 180; *see also* beams/girders, design considerations
fly ash 132, 180

glued laminated timber (glulam) 94, 181

Hammurabi's Code 4
Hooke's Law 16, 181
hydration 130, 181

I-joists (wood) 93, 187
International Building Code (IBC) 5, 181
International Code Council (ICC) 5
iron ore 45

k-factor 41, 182

laminated strand lumber (LSL) 94, 182
laminated veneer lumber (LVL) 94, 182
lateral resistance system 182
limit-state design principles 8, 182
Load and Resistance Factor Design (LRFD) 10–11, 13, 182
load combinations 7–8, 32–34, 182
load factor 8, 182
loads 6–7; dead/live/environmental 6, 179, 182; eccentric 179; gravity/lateral 181, 182; service/factored/design 8, 180, 185; static/dynamic 7, 179, 185
lumber *see* wood

mass timber 94, 182
micro silica *see* silica-fume
modulus of elasticity: for concrete 133; in deflection formula 38, 218–221; general 16, 183; for steel 49, 140; for wood 89
moment: definition 30; designing for 35; diagrams and formulae 218–221
moment of inertia 183; in beam deflection formula 38, 218–221; in column buckling formulae 42

nail laminated timber (NLT) 95
National Design Specification (NDS) Package for Wood Construction 5, 205–211
necking (of steel) 19, 183
neutral axis: of a concrete beam 135, 183; elastic/plastic 28

parallel strand lumber (PSL) 94, 183
permanent set 17, 183
pig iron 45, 181
plastic behavior 23–24, 26–27
plastic design 27, 183
plastic hinge 26, 184
plastic moment 26, 184
plastic range (for steel) 16–19, 184
plastic section modulus *see* section modulus
Portland cement 129, 184
pozzolans 132, 184
proportional limit 16, 184; for concrete 21; for steel 17; for wood 20

racking 185
radius of gyration 41–42, 184
rebar *see* concrete, steel reinforcing
rebar properties (table of) 159
reference design value *see* wood
resistance factor 10, 184

safety factor 7, 10, 11, 185
section modulus 27–28, 185
serviceability 8, 185
service load *see* loads
shape factor 29, 185
shear: in a concrete beam 143–147; diagrams and formulae 218–221, 185; general 36, 185; in a steel W beam 50; in a wood beam 90–91
silica-fume 132, 185
slag 132, 185
slenderness ratio (of a column) 41, 185; concrete 150; steel 52; wood 91
standard column curve 42, 52
static equilibrium 185
steel 44–53, 185; billets 46, 185; cold formed 47, 178; design considerations (beams 49–51; columns 51–53); design example 54–82; grades/strength 48–49; hot rolled 46–47, 181; manufacture/materials 44–46
stiffness 186; material 16
stirrups 145–146, 161–163, 168–170, 186
strain 15, 186
strain hardening range (for steel) 17, 19, 186
strength: allowable/available/design/nominal/required 8, 9, 177, 178, 179, 183, 185, 186; for concrete 20; material 15; for wood 19
strength design 12, 13, 186
strength reduction factor 12, 186
stress 14, 186; actual/allowable 9; compressive, shear, tensile 14; yield 8
stress distribution diagram 186
stress-strain curves 15, 16, 186; for concrete 20–21; for steel 16–19; for wood 19–20
structural composite lumber (SCL) 94, 186

tension controlled concrete sections *see* concrete
torque 187
torsion 187

ultimate strength 19, 20, 21, 187

water-cement ratio 130
web buckling/crippling/stiffening (of steel beams) 51

Whitney block 138–139, 187
wood (engineered) 92–95, 180; design example 113–127
wood (sawn): adjusted design values 87, 177; adjustment factors 87–88, 177; allowable stress 87; cellular structure/grain orientation 85, 181; design considerations (beams 89–91; columns 91–92); design example 96–112; reference design values 87, 184; species/grading 85; terminology 84
Working Stress Design 10, 12, 187

yield point 15, 18, 187
yield stress 8, 187
Young's modulus *see* modulus of elasticity